ARTIFICIAL INTELLIGENCE IN THERMAL SYSTEMS DESIGN: CONCEPTS AND APPLICATIONS

ARTIFICIAL INTELLIGENCE IN THERMAL SYSTEMS DESIGN: CONCEPTS AND APPLICATIONS

ENRICO SCIUBBA AND ROBERTO MELLI

Nova Science Publishers, Inc.
Commack, New York

Editorial Production: Susan Boriotti
Office Manager: Annette Hellinger
Graphics: Frank Grucci and John T'Lustachowski
Information Editor: Tatiana Shohov
Book Production: Donna Dennis, Patrick Davin, Christine Mathosian, Tammy Sauter and Diane Sharp
Circulation: Maryanne Schmidt
Marketing/Sales: Cathy DeGregory

Library of Congress Cataloging-in-Publication Data

Sciubba, E. (Enrico)
Artificial intelligence in thermal systems design: concepts and applications /Enrico Sciubba, Roberto Melli.
p. cm.
ISBN 1-56072-599-0
1. Heat engineering--Data Processing. 2. Computer-aided design. 3. Artificial intelligence. I. Melli, Roberto. II. Title.
TJ260.S34 1998 98-35551
621.4'00285'63--dc21 CIP

6080 Jericho Turnpike, Suite 207
Commack, New York 11725
Tele. 516-499-3103 Fax 516-499-3146
e-mail: Novascience@earthlink.net
e-mail: Novascil@aol.com
Web Site: http://www.nexusworld.com/nova

Printed in the United States of America

CONTENTS

CHAPTER 1

INTRODUCTION

1.0 - SCOPE OF THIS BOOK

This Book deals with the application of a powerful set of techniques, collectively denominated *Artificial Intelligence* ("AI" in the following), to general problems in the fields of the design, monitoring and control of thermal systems for energy conversion, with a heavier accent on design applications.

By necessity and definition, both fundamental and specialistic issues are beyond the scope of this essay: nevertheless, it is necessary that the reader (who is assumed to be conversant with Process Engineering but not necessarily with AI concepts and methods) be prompted on the one side to make his own judgment of the essence and of the implications of AI, and on the other side be allowed to have access to some *basic knowledge* about AI, to enable him to become an *intelligent* (in the "human" sense!) *user*, and to properly apply AI techniques by fully exploiting their power and recognizing their limitations. By no chance, both authors are *AI users* with Mechanical Engineering backgrounds, and have directly experienced the inherent difficulty of approaching a design or a control - or monitoring problem with *a basically non-numerical tool*, so different in principle from the exact deterministic "procedural" reasoning (IF...THEN...ELSE...) engineers are accustomed to. There are, of course, no unbiased authors: but, in our case, no effort has been made to lessen our biasing. Our belief is that the "user's approach" taken here is the fastest, most profitable and most efficient way to become a problem-oriented AI practitioner. Readers interested in more fundamental AI issues are referred to specialistic monographs [3,7,27,30,32,50,74].

This Introduction gives some indications of the *possible* applications of AI in the area of the design and operation of thermal powerplants, and a tentative list of the *actually implemented* applications: the fact that the latter is much longer than the former is a sure symptom that the field of "AI applications to Process Engineering" is in its infancy. Since AI techniques have been successfully and extensively employed in fields other than powerplant design, it is likely that the topics covered in this book will enjoy a rapid development in the near future. The dissemination of AI-related concepts is indeed one of the scopes of this essay.

1.1 - ENGINEERING DESIGN, KNOWLEDGE AND ARTIFICIAL INTELLIGENCE

From the early technological societies (onset of the Iron Age, approximately 3000 b.C.) to modern-day post-industrial society, the way the "engineer" has been perceived by the other members of his social group has changed substantially: early technologists (mainly, architects and blacksmiths) were considered little less than sorcerers, because they were able to master phenomena whose laws were not in the grasp of the rest of the community[1]. Later (in China, Greece and Rome, approximately from 500 b.C. to 500 a.C), the "engineer" started being considered an expert in a specific field (architects were not able to forge swords or assemble pumping devices, and vice-versa, steel makers had no knowledge of building techniques), and the design activity was seen as a mix of proper instruction (in fact, engineers often added their teacher's name to their own, as a patronymic) and of "a touch of genius". In the Middle Age, things rolled towards the Dark Side again: since - for a series of reasons - the laws of nature were not seen as a coherent set of causal relationships, but rather as an obscure body of imperscrutable metaphysical connections emanating by some transcendent mind whose thoughts humans could not comprehend, technologists became "sorcerers" again, and metallurgical, chemical, mechanical and civil "engineers" became alchemists, "masters of the sword" and "master masons". It is interesting to remark that these “specialists” considered themselves as “members-at-large” of a congregation or guild (one for each specialization), regarding themselves as a group of selected craftsmen endowed with (and striving to perfect) a certain very specific body of knowledge, which was not known to others, was essential for the correct performance of their work, and was to be communicated only veiled underneath a complex symbolic language that resulted obscure and unintelligible to non-members. This craft, and its symbolic systematization, had to be learned through years of apprenticeship: before being “licensed” as a “free master”, the former novice had to pass a series of formal tests to prove his knowledge of the “art”, and also to perform what we would now call an “experimental thesis”, by producing an artifact in accordance with the standards of his craft (the “masterpiece”).

[1] In reality, for all we know from the scant reports we have of those times, most frequently these laws were actually unknown as such to these early technologists as well! These ingenious craftsmen produced their artifacts on the basis of experience, common sense, heuristics and “intuition”. Trial-and-error was without any doubt the most common “design procedure” of those times.

After Galilei, Leibnitz, Newton, and especially since the Enlightenment age (approx. 1700), the terms "design" and "engineer" slowly went on to recover their pristine meaning which they still carry today: in the last decades, the effort to streamline and rationalize the design activities has fostered the development and the implementation of highly integrated team-work techniques in the engineering profession. What we have inherited from the dark years is the insistence in adopting a particular form of a highly specialized "technical" jargon, that is in fact equivalent to a symbolic pseudo-language, that can be useful to make technical statements less lengthy and more immediately understandable by the expert, but that at times veils by its excessively formalized schemes a simple underlying phenomenology, making it difficult to establish connections between similar physical facts that are expressed by seemingly different formulas.

As of today, it seems that different Authors [1,5,8,12,65] have finally reached the consensus that Engineering Design is a *highly structured, interdisciplinary, creative process* which is invariably the result of a tight cooperation among several individuals (who formally or informally constitute the *Design Team*) whose diverse expertise and specialistic skills must be blended together and result in an Engineering Proposal (consisting of drawings, calculations, and all necessary technical documents) about a *new solution to the problem which prompted the team into action.* In the above, the words "creative" and "new" have been underlined to reiterate the concept that *the product of the design activity is always an original answer to a technical, economic, social or marketing problem*: naturally, past experience (both general and specific) plays an important role in the formulation of the solution (as an example, consider the striking similarity between gas turbine plants, or Diesel engines, or airplanes, manufactured by different producers), but "design" brings with itself a flavor of "innovation".

In analyzing the design activity as a whole, the concept which first comes to mind is *knowledge*. To originate the need for a design, knowledge about a (potential) specific problem is required; to initiate the design activity, general knowledge of the related engineering fields is needed; to participate to a design project, specific knowledge in a particular field is mandatory; to successfully perform as a design team, team members must be able to suitably communicate their knowledge to each other; to transport the design from paper into reality (i.e., to build a component or a machine or a plant) the team has to impart this knowledge (or at least a portion thereof) to Production Engineers; to instruct operators and final users on how to make proper use of the designed item, some other form of knowledge must be communicated, and so on. So, one should consider - as an engineer - that a design activity, besides its *quantitative* side, has an equally important *qualitative* side; that, in addition to performing engineering calculations, it is necessary to be able to *communicate to third parties not only the results of these calculations, but also the original problem definition and the method which was used to reach a solution*; that, in end effect, *to obtain a better chance of future design improvements, it is more important to leave written record of the* **why** *and of the* **how** *the design goal was achieved starting from certain premises, than to give an exact description of the numerical calculations which led to the definition of certain design parameters.* We shall return to these concepts in Chapter 2. Here, it suffices to say that Artificial Intelligence techniques serve precisely this purpose: they can be used to describe in qualitative terms the logical steps of an engineering activity, thus formalizing and systematizing a large body of knowledge, and assisting engineers by offering them logical guidance and support in their design endeavors.

1.2 - WHY ARTIFICIAL INTELLIGENCE?

Artificial Intelligence (for an "exact" definition see Sect. 3.1.1) is in reality a cumulative denomination for a large body of techniques which have two general common traits: they are *computer methods* and they *try to reproduce non-quantitative human thought processes.* For design applications, there is a third common trait: *problems handled by AI-based techniques are usually ill-structured*, i.e., they cannot be tackled with pre-determined solution models. See also Section 2.1 for a better definition of ill-structured problems.

There are many astonishing applications which have enjoyed much popularity among non-specialists, like those based on pattern, image or voice recognition, to mention a few. The applications we will deal with in this essay are related to a smaller subset of general AI techniques: the so-called *Knowledge Based Systems*, also called *Expert Systems*. Referring the reader again to Chapter 3 for the adequate definitions, an expert system ("ES") can be roughly described here as a *specialized AI application aimed at the resolution of a single and well-identified class of problems*. It is almost not necessary to remind the reader here that neither a computer nor an ES can "think": the ES is a sort of well-organized and well cross-referenced "task list", and the computer is just a worktool. Nevertheless, ES (and in general AI techniques) can result in very efficient, reliable and powerful engineering tools, and can do to *qualitative* Engineering just as much as numerical methods have done to *quantitative* Engineering. Some of the major benefits offered by expert systems are:

- they provide an efficient method for encapsulating and storing knowledge so that it becomes an asset for the ES-user. For example, after an ES has been built and fielded by a Company to assist its Design Engineers in the choice, design and technical specifications of a steam boiler, all future designs will be carried out in a much shorter time and with a less intensive use of both high-level human resources and computational hardware;
- they can make knowledge more widely available, and help overcoming shortages of expertise. So, if an ES is developed by the R&D Division of a tire producing Company to overview the adaptive control of the pre-vulcanizing and vulcanizing processes, it can be applied in all of the factories of that Company, even in remote locations, as long as the process

remains in use, without the need of devoting highly qualified human resources to the proper communication of the technology;

- knowledge stored in an ES is not lost when experts are no longer available. This is for instance the case of a numerical process simulator, developed and implemented in a machine language which then becomes obsolete. If an ES (which is language-independent) has been employed in the development of the simulator, then much fewer high-level human resources must be employed to re-implement the code in a different language or under a different operating system environment;
- an ES stores knowledge it can gather from experts in some field, even if they are not computer specialists (and even if they are no computer users!). For example, when prospecting for new raw resources, it is often indispensable to gather non-specific knowledge about the nature of the area to be scouted, or about the type and history of possible ores or flows that may have surfaced in the past. When performing this "information gathering" in remote areas, an ES- assistant that prompts, collects and critically analyzes answers by natives can drastically reduce both the time needed to make an educated decision and the amount of resources invested in the prospecting activities;
- an ES can perform tasks which would be very difficult to perform with more conventional software: thus, it becomes possible to implement features which were impossible or very cumbersome with earlier systems. As we shall see below, in Chapter 3, ES's allow the handling of approximate knowledge (which cannot be hard-coded otherwise) with relative ease. For example, while designing a Heat Exchangers Network ("HEN"), it is at times convenient to relax the rule " no heat flux across the pinch", and reformulate it as "*some* small measure of heat flux across the pinch is allowed, provided the hot or cold utility are *conveniently* reduced by this measure". This construct would be impossible to translate into any rigidly-framed type of computer instructions, but can be easily managed by an ES, that treats the word-concepts "some" and "conveniently" in a *relative* way, just as we mean it when we verbally formulate the relaxed rule.

Other practical advantages associated with the development of a knowledge based system are:

- rapid prototyping
- explicit knowledge encoding
- possibility of dynamic and efficient further developments
- ease of alteration
- explanations capabilities for development and validation
- potential for rapid inclusion of new rules and techniques

All of these benefits come at a price: high requirements in computer resources, and low speed of computation. The amount of physical memory space required by a Knowledge Base is very high because qualitative knowledge is not easily amenable to a simple numeric (binary) representation. And, since the number of operations per clock tick is of course the same as in "numerical" codes, but qualitative rules are implemented in machine language by complex combinations of elementary logical operations, the actual execution time of a "rule" is much higher than that of a numerical operation. Furthermore, developing and installing a large ES is a high added-value operation, and therefore a very expensive enterprise: the provisional returns should be always carefully scrutinized before launching an ES-project. There are two main factors which can make knowledge-based systems applications profitable:

1. the specialistic work load could be - or become - higher than that which can be reasonably performed by the available experts; or, some of these experts might be no longer available in the foreseeable future;
2. the existing - or foreseen - working conditions could lead to too high a demand on decision-making engineers; this, in turn, may result in what goes under the name of *cognitive overloading* (which in plain English means simply "too many facts to analyze in too short of a time before making a decision"). When cognitive overload sets in, not even increasing the number of experts dedicated to the problem might solve it, because communication among experts may deteriorate (too many information bits of very different kind must be communicated to an increasing number of people).

Before the reader proceeds any further in this Book, it is useful to present here our confutation to some of the more common misconceptions about AI in general and ES in particular. The list has been adapted from [27].

1. *It is not true that ES's are suitable for every type of design problems*: for instance, a very large class of structural, thermal and fluid dynamic problems can be tackled very effectively with "exact" (and fully deterministic) algorithmic techniques, and rather poorly with ES. On the other hand, it is not correct to describe ES as "an alternative to conventional computer programming": in the great majority of practical design cases, the fields of application display a substantial overlap.
2. *It is not true that AI-users must be conversant with AI programming languages*. Neither is true, as a matter of fact, that creators of ES must be familiar with high-level languages like PROLOG, LISP, and the like: actually, often a sound knowledge of the principles of predicate logic, complemented by specific domain knowledge, is all it takes to develop an effective ES.
3. An ES User, however sophisticated and knowledgeable in his field of expertise, is logically, functionally and mentally distinct from a Knowledge Engineer. Similarly, a Domain Expert is a substantially different professional figure than a Knowledge Engineer.
4. In turn, not every Design - or Process Engineer or Technical Operator is a Domain Expert: actually, there is a remarkable scarcity of reliable and

actually knowledgeable Domain Experts in any field (including Sports and Politics!).

5. On the other hand, an ES should not be tested "against the Domain Expert", i.e., its performance ought not be compared with that of the human expert in a similar situation.
6. Complete modularity is rarely achieved in logically complex ES, and even in "small" ES's it is not possible to simply add to (or delete from) the Knowledge Base groups of Rules and Objects, or even a single Rule or a single Object, and be certain *a priori* that the solution or the inference process are not affected.

1.3 - "POSSIBLE" VS. "EXISTING" APPLICATIONS OF AI TO THERMAL SYSTEMS

Basically, an AI procedure is capable of reproducing a series of actions which would be taken by a human "expert" (or team of experts) in response to a certain stimulus. In the field of design and operation of thermal plants, it is possible to recognize that most actions taken by the engineer or by the plant operator are amenable to codification: indeed, design textbooks and manuals, as well as operating manuals, are *just a codification of what past experience has taught us with regard to certain specific tasks or events*. This "experience" goes well beyond simple instructions as "for an operating temperature T, use bolts of such and such material": most designers will agree that there is a "feel" for design which can be gained with specific experience, or inferred by intelligent comparison with similar cases, or communicated by an expert in the field (a teacher, a senior engineer...). Similarly, plant operators gain their knowledge by becoming familiar with specific procedures (manuals), or by specific experience "on the field", or by being taught by experts (senior operators or design engineers). What is of importance here is that *this body of "knowledge" is - or can be - codified in such a way that it can be conceivably communicated to a computer*. The implications are clear: the computer can replace or supplement the human designer/operator and, being the former much faster, memory-intensive, even-performing and reliable than the latter, the design or the monitoring process will gain from this "computerization". According to this point of view, virtually all engineering tasks could be performed with the help of (or entirely by) a properly programmed computer. A list of the possible specific fields of application of these AI techniques (listed here in increasing level of complexity) is given below:

- Process control (chemical, thermo-mechanical)
- Maintenance management (in real time)
- Emergency procedures
- Production scheduling (in real time)
- Components design

- Process design and optimization
- Qualitative process modeling
- *A priori* thermoeconomic optimization

By contrast, the list of the AI applications actually implemented for the above tasks is not as extensive as one would expect:

- Several applications in *production scheduling* (most of them not-in-real time)
- Several applications in *process control*, to assist or replace the operator in normal plant-conducting mode (notably, in emission monitoring);
- Some applications in process monitoring (mostly in the chemical and nuclear industry)
- Some on-line maintenance management procedures (notably, in the aircraft maintenance field)
- Very few emergency-handling procedures (mostly in the chemical and nuclear industry)
- Several applications in simulators for training purposes
- Some applications to prototype testing, including automatic procedures for data reduction/interpretation and for intelligent "state checking"
- Very few design procedures, the majority of which performs a sort of "database scanning" activity to select the most proper component from a fixed, pre-assigned list of possibilities
- Very few expert, almost-real-time cycle- and plant simulators
- Virtually no qualitative modeling, because the tendency is that of having the computer do the "number crunching" job, so that a group of experts (the "Engineering Team") can take their decisions exercising their judgment on the results of the simulations
- No Thermoeconomic Optimization. Actually, there are some AI applications in the field of economics, but none of them has been translated into a suitable engineering equivalent.

The causes of this discrepancy between what could be done and what is actually implemented at the application level are diverse: there are of course technical reasons, like the lack of good programmers and the intrinsic computational intensity of AI techniques (that require, more than other computational techniques, large amounts of physical resources, etc.), but they are not substantial, and cannot be invoked to justify such a wide gap between theory and practice. A more fundamental reason is that AI is a relatively new field, and that its theoretical foundations are still being discussed, so that we are still in the "run-in" phase of the technology. Another reason is a certain mistrust expressed by senior designers in a tool that - consciously or unconsciously! - some of them feel will "take their place": as a consequence of this general attitude, there have been difficulties in getting "experts" of a specific field volunteer in transmitting their knowledge to a computer code ("Knowledge Gathering Problem", see Section 4.2); there has been a sort of passive opposition on the part of plant managers to substitute an Expert Monitoring System for their reliable “hybrid” systems constituted by low-level control- and monitoring devices governed by well-trained and experienced Human Operators; there have been problems in convincing Design Teams to work together with Knowledge Engineers in developing what in effect they see as their electronic duplicate, and so on. The technical shortcomings will be disposed of in the immediate future, by the fast pace of the development both in the hardware and in the software. The second problem, constituted by the “immaturity” of AI techniques, will gradually be overcome by advances in the theoretical field that will hopefully lead to the construction of a complete, formally well-defined, coherent and unified “Applied Cognitive Science”. The mistrust on the part of engineers and the reluctance of plant managers will probably disappear after one natural generational turn-over: since the adoption of a new technique depends essentially on the degree of substantial comprehension and on the amount of “feeling” for it that its (potential) users have developed, this is clearly an educational problem, and will gradually be overcome when new generations of engineers can avail themselves of new generations of codes they know they can rely upon.

1.4 - A WORD OF WARNING

Before the reader proceeds further through this Book, we feel that a word of caution is in order: this monograph, in spite of its apparent formal "exactness", is indeed - as it will become apparent in the course of the following Chapters - a rather introductory-level, descriptive compilation of AI concepts and possible applications. We do not anticipate (and the reader should not expect) that, upon completing the study of the topics contained here, non-AI-specialist engineers will be able to write their own AI application. To use a term which will be frequently encountered in the following pages, *much more specific knowledge* is needed to be able of writing successful (and even unsuccessful!) applications. In particular, notions of Formal and Symbolic Logic, of Predicate and Propositional Calculus, and of the specific Relational- or Object-Oriented languages are needed. The most a reader can hope to become after digesting the concepts and the examples discussed in this Book is an *informed and competent* user. For those interested in specialistic AI monographs, Ref. [26,34,35,74] are recommended.

CHAPTER 2

THE DESIGN OF THERMAL SYSTEMS

In this Chapter we describe and study in detail the general activities connected with the Design of Thermal Systems, which is the only design problem of our concern in this Book. The problem position will be discussed first (Sect. 2.1 and 2.2), to analyze the various forms under which a design task can present itself to an engineer. Then, a set of general design guidelines will be developed and discussed: these guidelines are only in small part original, and for the major part are derived from either Design Manuals or textbooks or otherwise published and accepted Design Procedures. The important point that we want to make here is that there is indeed *one* general design protocol for all types of design problems: this point is crucial to the thesis developed throughout the Book, because it is in fact the justification for our search for AI-based "Design Assistants". In Section 2.3, we show that the two fundamental design tasks encountered by an engineer, namely direct (simulation) and inverse (design) problem, are amenable to a single meta-procedure. Section 2.4 contains a brief conceptual description of the two types of computer-aided design tools available at the present state of technology: process (or plant-, or system-)[1] simulators, which are computer codes capable of calculating proper numerical values for all design parameters of interest, and process (or system-) "synthesizers", which are Expert Design Assistants. Optimization forms the subject of Section 2.5 and 2.6. Our claim here is twofold: first, we will show that there are two levels of optimization, one being the meta-level of the other, and second, that a true optimization of Thermal Systems can indeed be based on thermodynamic

---

[1] A *process* here means a perfectly identified series of mechanical, chemical and thermal transformations that modify the thermodynamic state of some well-specified form and quantity of matter, possibly releasing or absorbing some energy flux in the course of the transformation. A *system* is an interconnected set of components whose goal may or may not be that of realizing a process. For example, a combined-cycle power plant constitutes, in the sense described here, a process; but it may also be regarded as a system. Any organized human structure (an industrial complex, a city, a highway, an information network) constitutes an (engineering) system, but it cannot be regarded as a process.

quantities alone, provided the representative function is expressed in terms of exergy instead of energy. We will show also that all Thermoeconomic optimization techniques can be included as sub-cases of the Exergetic optimization method.

2.1 - IS THERE A "UNIVERSAL" DESIGN PARADIGM?

Before we embark in a discussion about the possibility of implementing "Expert Design Assistants", we must answer a fundamental question: is it possible to construct a "universal" design paradigm that is so omni-comprehensive to describe every conceivable act of design?[2]

To answer this question, we must examine what a design task consists of. A little reflection based on every Design Engineer's experience shows that an essential feature of most design tasks is that *they are posed as ill-structured problems*. An ill-structured problem is one that [47]:

1. cannot be described solely in terms of numerical variables, scalar and vector quantities;
1. possesses goals that cannot be specified in terms of a well-defined objective function;
1. admits no algorithmic solution.

Ill-structured problems are often called ill-defined, dynamic, or uncontrollable [74]; but their most striking feature is that their solutions are *unpredictable*, in the sense that the environment in which the solution is to be sought for has a strong influence on the existence, unicity and type of solution. To solve these problems, an engineer relies on judgment, experience, heuristics, intuition and analogy rather than on specific knowledge of solution procedures applicable *per se*. Actually, the question whether ill-structured problems can be solved by some kind of structured engineering reasoning has been posed long ago, and it has been answered in the affirmative, in direct and indirect ways [1,4,76]: otherwise, there would be little motivation for this book! It is useful though to reconsider our conception of a "design activity", because there are two concepts that must be clearly understood by an engineer: first, that what goes under the name of "design activity" is in reality a concatenation of several complex actions, only some of which fall within the responsibilities of the engineer. Second, that if we compile a detailed list of the various sub-activities that constitute a "design procedure", it is indeed immediately apparent that all of these activities are common to (i.e., they must be performed in the course of)

---

[2] *A paradigm is*, paraphrasing [74], *a common principle of organization which constitutes the conceptual basis for a programming language or a solution procedure.* Notice that a paradigm is not necessarily algorithmic: as a matter of fact, we shall see that most AI applications to the field of Thermal Design are by force non-algorithmic.

almost every design task. There may be exceptions, but it is intuitively clear -and can be demonstrated by a comparative analysis of a large number of published design procedures- that these exceptions are caused by peculiar characteristics of the specific project, and in no way subtract to the validity of the "general design procedure", of which a logical flowchart is shown in Fig. 2.1. We shall proceed to a brief critical review of this flowchart (Section 2.1.1) before we analyze the main Design Phases individually (Sections 2.2÷2.4).

After we convince ourselves that *every* Design Procedure is composed of the same building blocks, we can also strengthen our intuitive idea that there must be a universal underlying paradigm in the solution of every engineering task: though there is no proof that this is indeed the case, no proof to the contrary has been found yet. So, we shall assume that it is always possible to devise a specific instance of a (in principle yet unknown!) "Universal Design Paradigm" that results in the successful solution of a specific instantiation of the Universal Design Procedure described in the following Section.

2.1.1 - THE "UNIVERSAL DESIGN PROCEDURE": A POSSIBLE FLOWCHART

In the real world, a design project includes both technical and non-technical tasks: often, the physical "design" (if we take this word in its restrictive meaning of "quantitative sizing of units and systems") is a minor activity in the general project perspective. Several other activities are of importance before the "sizing", during (i.e., concurrently with) it and after its conclusion, and it is their coordinated sequence that constitutes the actual "design" task and is accordingly managed in its totality. Extrapolating from Design Manuals and textbooks, a "Universal Design Procedure" can be identified and is presented in Fig. 2.1. With reference to this flowchart, let us examine in detail the single activities.

2.1.1.1 - Definition of Needs and Objectives

Scope of this task is to explicitly express the exact and complete explanation of the reasons which justify the investment of resources in a specific Project. The "needs", as well as the "objectives" of a project are not always formulated in terms of economic convenience: often, social equity requirements, social opportunity, macro-economics or even political interests can be properly used as criteria to include in the description of needs and objectives. Generally speaking, this Phase of the Project pertains to the "Management"; some fairly recent developments consist in the introduction of a sort of systematic feedback mechanism in the hierarchic structure of Companies and Agencies that allow for some "input from below", i.e., for the possibility that suggestions from the basis of the organizational pyramid (typically, blue collar Workers) can reach, through a proper filter constituted by intermediate Management, the upper decision-making levels. In larger Companies or Agencies, another structure which can functionally cooperate in the definition of needs and objectives is the R&D Department.

2.1.1.2 - Definition of the Criteria for Success

Scope of this task is that of identifying one or more quantitative attributes of the product to be designed, in such a way that the numerical values of these parameters can be computed in advance (albeit approximately) and can be used to determine the preliminary degree of success of the Project. Although the most common parameters are those linked to economic considerations (like "number of sales" and "gross unit profit"), social, political and opportunistic criteria are also routinely employed.

2.1.1.3 - Assessment of the Probability of Success

This is a very difficult task, because at this stage of the Project, the assessment of the probability that the product generated by the design activity is indeed produced according to its specifications and schedule, meets the objectives, satisfies the needs, and corresponds to the criteria for success is perforce based on estimates and educated intuition. This activity must therefore be performed by experts in the specific technological or industrial niche, and the results ought to be tightly monitored in the following of the design activities.

2.1.1.4 - Market Analysis

This task consists of marketing studies, sales forecasts, preliminary profit assessment, etc. The participation of engineers to the activities of this Phase is usually very limited or absent altogether: experience shows that often, on the contrary, a deeper involvement of Process and System engineers in this task allows for a better interface between marketing and technical people, and improves both product definition and the engineer's "feel" for the relative importance of the various design parameters.

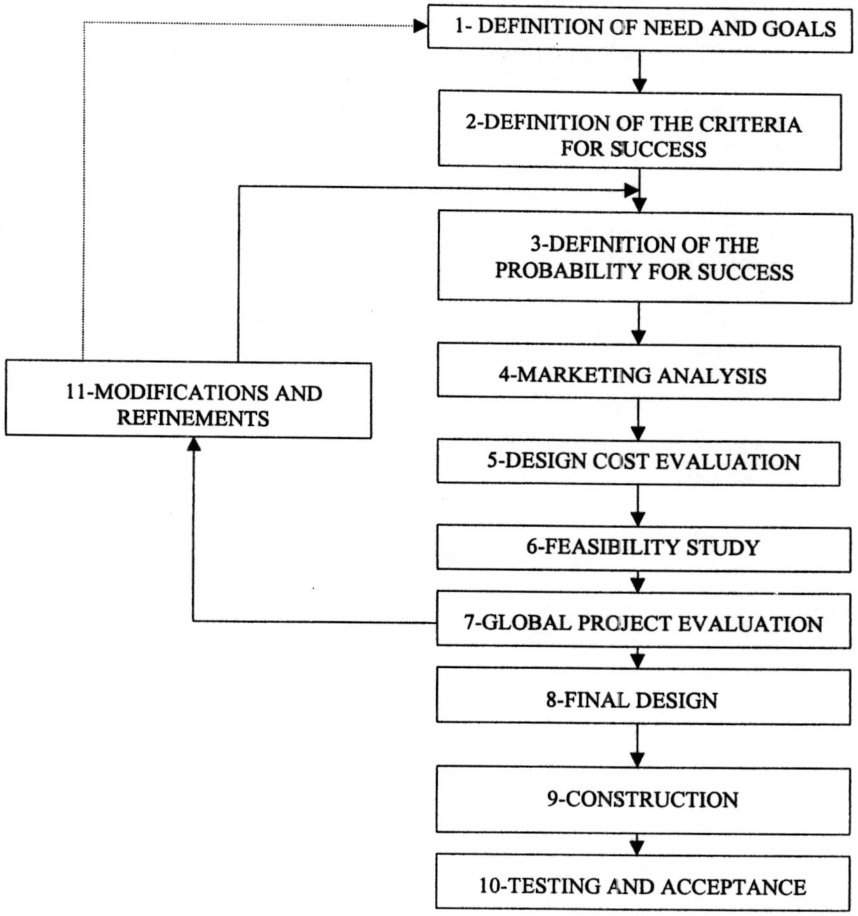

Figure 2-1. Universal Design Procedure

2.1.1.5 - Preliminary estimate of the design costs

This task is usually performed by a specific Division of the Company or of the Agency which is responsible for the Project Management (the Division responsible for Technological Methods and Production Scheduling), on the basis of detailed specific information provided by the Technical and Commercial Divisions interested in the Project. Errors made in this Phase can be very costly: since the specific design problems are usually unknown in detail at this stage, care must be exercised on the one side to limit the risk of cost overruns (which have been known in some cases to reach values in excess of 1000% of the original estimate!), and on the other side to avoid overestimating the costs, which could possibly result in the preliminary rejection of a lucrative Project.

2.1.1.6 - Feasibility Study

This is a well-codified engineering activity, aimed at determining whether all conditions which make the Project feasible are met at the time of its foreseen realization and in the operative conditions under which the Project will be undertaken. To be feasible, a Project must be:

- technically possible
- operationally reliable
- industrially sustainable
- economically advantageous
- legally acceptable

2.1.1.7 - Overall Evaluation

This is a task for upper Management: it is a strategic decision which is often taken after a formal presentation of the Project made by Design and Marketing experts.

2.1.1.8 - Final Design

This is what is usually called "design": it is a very complex activity, in which the item, unit or system to be built is completely designed to specifications, by means of a concerted cooperation of process-, structural-, industrial-, mechanical-, chemical-, material-, environmental- and control engineers, possibly helped by field experts for specific problems.

2.1.1.9 - Construction

This task is directly supervised by specialist mechanical-, chemical-, structural-, or civil engineers, who have specific knowledge and experience in the "construction" field (where by *construction* we mean all of the possible production technologies that lead from the raw materials to the end products).

2.1.1.10 - Testing and customer's acceptance

This task is performed by Quality Assurance specialists, with the assistance of Design engineers. Usually, two teams work jointly on this activity: the Quality Assurance (Q&A) team of the Constructor and that of the Customer. If required by the contract, a third independent party can coordinate the activities of the first two ("Arbitrate").

2.1.1.11 - Modifications and improvements

This is one of the activities of the R&D Division, but requests for modifications are mostly originated by the Process Engineers responsible for the maintenance and operation of the hardware. Very useful for improving the technological production line, this activity can potentially become dangerously wasteful if too many unjustified modifications are proposed, because the amount of resources which must be allocated for the analysis of each unsuccessful process modification is almost as high as -say- that required by the Feasibility Study.

2.2 - Application of the Universal Design Procedure to Process Engineering Problems

2.2.1 - Formulation and Position of a Process Engineering Design Task

An Engineering task pertaining to the (partial or global) design of a Thermal System is, in a logical sense, a problem generally amenable to a meta-mathematical formulation, which consists of a set of *goals*, a set of *premises* and a set of *constraints*. The goals are given under the form of qualitative or quantitative descriptions of the "products" of the system to be designed: they can be physical goals in a strict sense (tons of steam per hour at certain thermodynamic conditions, Megawatts of installed power, liters per minute of desalinated water, etc.), or less immediately physical, but unequivocally engineering-like specifications (global efficiency of a compressor under certain well-specified conditions, overall performance of a Heat Exchanger Network, a specified value of stack emissions of certain pollutants, etc.). The premises, also called the *design data*, consist of quantitative or qualitative information about the available inputs the engineer can use to achieve the goals (quantity and chemical composition of the available cooling water, type, chemical and physical properties of participating media, including fuels and raw materials, information about the immediate surroundings of the proposed plant, etc.). The constraints are usually given under the form of a list of affirmative and negative propositions about either the outputs of the process and their physical impact on its immediate surroundings, or about limitations imposed on inputs, outputs or on the physical structure of the plant (pollution limitations, visual impact, environmental regulations, maximum or minimum price of either the equipment or the unit of product, social- or economic limitations imposed by laws and regulations, etc.). Ideally, an Engineering Task is always well-posed: i.e., the inputs are completely and unequivocally specified, the outputs exactly quantified, and the constraints exactly known. This is the type of Engineering Design problem which is usually described in textbooks, and in fact often goes under the name of "textbook problem". Such problems always admit one or more perfectly defined quantitative solutions, i.e., they can be treated using well-defined design procedures to proceed from the premises to the goals taking into account all constraints, and correspondingly result in at least one, and usually more than one, process- or system structures consisting of a certain number of well-identified components which act on the input fluxes of matter and energy to produce the required outputs of matter and energy. There can be cases in which there is no possible solution, but they correspond to easily identified situations in which the state-of-the-art of that particular branch of technology does not allow the required input→output transformation because of limitations usually related to the performance of one or more of the pieces of equipment necessary for the realization of the process. This would be the case, for instance, of a fuel-to-power transformation with an overall efficiency higher than 50%, if the fuel is natural gas and the cooling medium available is air at 313°K, or, to mention a problem widely discussed in the media today, of electricity generation based on nuclear fusion.

Unfortunately, real-world problems are seldom textbook problems: though the goals are usually well-defined, data are by force incomplete, and often expressed in a qualitative instead of quantitative fashion, and constraints are weak[3] or, worse yet, vague. These cases must be handled nevertheless by the design engineer, and they of course are, because processes are devised and plants built even under these conditions: but some of the comfortable certainties offered by textbook problems are no longer available. So, there is no *a priori* certainty of the existence of even one, leave alone several, quantitative solutions, because generally there is no well-defined *specific* design procedure the engineer can follow to proceed from the premises to the goals taking into account all constraints[4]. Consequently, while it is true that even in these cases there may be more than one process- or system structure consisting of a certain number of

[3] A weak constraint has the form "p not higher than C" or "q at least equal to C", i.e., "$p \leq C$" or "$q \geq C$". A strong constraint has always the form "$s = C$".

[4] This is not at odds with the thesis of this Book: while specific, i.e. case-dependent, particular design procedures may not be available for all types of design problems, it is always possible to formulate, tackle and solve an Engineering Design Task within the guidelines of the "General Design Procedure" outlined here.

well-identified components which act on the input fluxes of matter and energy to produce the required outputs of matter and energy, the type and performance of these structures depend on some choices the engineer must make along the way, and the results must be checked thoroughly against the specifications, to make sure that only the available inputs are indeed necessary, that all of the required outputs are indeed produced, and that all the imposed constraints have been correctly abided by.

This calls for a redefinition of the Engineering design activity: since we are searching for a general design procedure, if we accept that most design tasks are ill-posed (from a logical-mathematical point of view, it would be more correct to say that they are not completely specified, or fuzzily posed), the first step in any design procedure is that of understanding the real problem which is hidden behind the fuzzy formulation. Some Authors (notably, [1]) call this incomplete formulation the *primitive problem*, but, since we will assume that all engineering problems are formulated this way, we will not need a special name for this, and reserve the name of "Textbook Problem" to the well-posed form of task specification, which we shall not treat at all, because its treatment will be implicitly embedded in our "general" design procedure as a special (easier) case. The premise from which the design engineer has to start is very simple: if (= since!) the problem is ill-posed, a well-posed form must be devised before a solution is attempted. The strategy for this task is at best described as follows:

1. The set of inputs must be examined and de-fuzzified. This means first of all that a preliminary investigation must be carried out to determine what are the actual inputs, what is their quantitative availability, what are their accepted chemical-physical properties, and what is their cost[5] as inputs, i.e., at the boundaries of the system which we are about to design.
2. The goals must be similarly analyzed, and possibly re-defined, so that they are clearly and completely identified: since it is often the case that a problem is over-specified (but still fuzzy!), it is useful to divide the goals in *mandatory* and *accessory* goals. Mandatory goals *must* be achieved, i.e., they must be included in the "criteria for success" (see Section 2.1.1.2), while accessory goals can be used as "weights" to assess the *desirability* of a solution, very much as "bonus points" which may make one of the solutions more attractive than another for both designer and customer. The identification of mandatory and accessory goals can have a large impact on the structure of the resulting process, so it is necessary to exercise prudent judgment in the choice, to make use of the advice of an expert of the field, and to seek at least a partial customer participation in this activity. It is also advisable to keep track of the

[5] Notice that this "cost" must not be necessarily expressed in monetary units. Though this issue is the matter of some debate [1, 16, 78], it is our firm opinion that the proper cost to attach to both a flux of raw material or a piece of equipment ought to be its *cumulative exergetic cost*, expressed in kJ/kg for the streams of matter and in kJ/(kW installed) for the single piece of equipment [21, 66, 77].

initial selection, and to be ready to modify it at a later time in the design procedure, should it appear that the initial (by its nature, somewhat arbitrary) choice is likely to have a negative impact on the final outcome.

3. Equally important is the correct interpretation of the constraints. The first rule (always reported as a strong suggestion in design manuals) is that of taking the safe route and impose weak constraints in a strong fashion. But if a constraint is formulated in a vague propositional sense, like for instance "taking due care that the efficiency of the compressor is the highest possible attainable under the given conditions", the design engineer must *interpret* this rule, and re-formulate it in a form which is immediately useful for design purposes, like "given that the stream to be compressed is air at certain (exactly specified) inlet conditions and with the design mass flow rate *m*, and that the required pressure ratio will be between 4 and 12, the multistage compressor shall be determined by the performance curve described in the design procedure, with a maximum acceptable deranging of -1% absolute". It appears clearly that this "constraint defuzzification" is a delicate task which can very much make the difference between a successful design and failure, and that it should be also performed under expert advice and discussed in advance with the customer.

2.2.2 - TOWARDS A GENERAL PROCESS DESIGN PARADIGM.

Once the problem position has been properly analyzed and, if necessary, re-formulated, the design engineer has already gone a long step in his path towards the solution. Naively, our primitive idea of a "general design paradigm" would suggest that, at this point, having reached a complete and comprehensive understanding of the goals he is asked to achieve and of the raw materials he can use, the design engineer performs a *creative act* of conceiving, critiquely revising, and developing to final design form one or more processes that can be used as *means* to proceed from the raw materials to the goals. We are faced with a semantic problem here: traditionally, the concept of *creativity* has been applied to acts of artistic or literary production, rather than to engineering activities. For these, we use instead the word *ingenuity*. In fact, there is no functional or mental difference between the creative act of -say- a painter and that of a process engineer: what is different in the two cases is the problem position! A work of art is originated by a *very fuzzy* problem position, in which especially the constraints and the goals have no need to be formulated in a precise formal way, but quite on the contrary are more effective the more "abstract" they are (nobody would ever object to a sentence like "The artist must express himself freely"). If we understand this, we see that our "naive" idea was not so naive after all: the design paradigm *must* indeed begin with some creative, intuitive "concept generation" [1]. Of course, the freedom which this engineering creativity enjoys in practice is not comparable to that of an act of art. In particular, there is a vaste body of knowledge ("experience") that engineers avail themselves of, and that sets strict bounds both on the type and number of possible choices: this limiting action is so severe that in most instances there are very few alternative routes open for this creativity. But, this is again an inessential fact, which depends on the relatively high level of technological development at which we consider our "process design": our present technical knowledge embodies in fact all acts of past creativity, and correctly so, because nobody wants to "reinvent the wheel"! This discussion may seem philosophical and removed from the realm of practical applications, but it is instead very practical in character, and it enables us in fact to proceed with the formulation of a general design paradigm which allows for a precise formalization of the design procedure and will form the basis for the subsequent application of AI techniques to process design. Synthetically, these steps can be described by the following list:

1. Consider all possible processes that can be conceivably devised to attain the goals from the premises; usually, it is convenient to proceed from the simplest (fewer components) to the more complex process, because the engineer's Knowledge Base is by its own nature additive: but this is not a necessary requirement. What is important is that these processes are listed in an orderly fashion, with a clear ranking expressed by a well identified parameter, regardless of their known "implementability" from past experiences.
2. Perform a detailed conceptual analysis of some of the configurations which seem more "promising", based on past experience, external advice (including the customer's) or educated engineering intuition. It is likely that by doing so, additional configurations may be discovered, which can be obtained by the previous ones by "perturbation", i.e., by adding or deleting or modifying one or more components, but keeping the general process structure unchanged: in particular, it is important that all configurations retained in the list at this point satisfy completely all mandatory goals. Review the list originally compiled, pruning the most complex or apparently less feasible configurations and adding some of the new ones. It is important to remark that at this level all choices are qualitative or, at the most, semi-quantitative, and that these two first steps are substantially based on the engineer's experience and creativity.
3. Screen the new list critically, with the stringent goal of reducing the alternatives to a small enough number that quantitative calculations can be performed on all of them. The screening criteria ought to be based essentially on the constraints contained in the problem formulation, and on any accessory constraint that the engineer may deem appropriate at this time (for instance, socio-economical or technological situation of the area in which the process will be implemented into a plant, or known availability of some components on site).
4. Having reduced the alternatives to a "small" number (how small this number actually is depends on the amount of resources that can be committed to this preliminary feasibility study), perform a preliminary calculation of the simplified process (neglecting, for instance, pressure losses in pipes and valves, secondary heat losses to surroundings, etc.). Compute efficiencies,

specific fuel consumption, and any other coefficient of performance that may be of interest; also, perform a preliminary sizing of the major, non-standard equipment. Discard all configurations which result in unusually over- or undersized components, which have exceedingly low efficiencies or low reliability, or which have a much larger environmental impact than the others: in practice, discard all configurations which have the least (defined in some sense) probability to achieve a desirable total of accessory goals.

5. Formulate an objective function (which may include, but ought not to be limited to, economic considerations) to assess the "goodness" of the process we are searching for. Perform a more refined process simulation, and compute this objective function at design and (if necessary) at off-design conditions. Select the few alternatives for which this objective function clearly attains its "optimal" values, and repeat the refined process simulation for each one of them, under realistic operational conditions (load-following transients, start-ups and shut-downs, varying external conditions). If enough data are available, perform a life-cycle analysis.
6. Select a final number or two or three alternatives and review them critically in detail together with the customer and any field expert you can find. Once the "final" configuration has been agreed upon, proceed with detailed design and operational numerical simulations.

This list applies equally well to the design of an automobile or to that of a water desalination plant. Naturally, there are substantial differences both in the technical details and in the technological state-of-the-art of the individual components, but these differences can be addressed internally to each step, and do not detract to the generality of the "main" procedure, which is, in effect, *a meta-procedure applicable to all Process Design activities*. The flowchart of Fig. 2.2 represents the various "tasks", each expanded in its sub-tasks for ease of comprehension: notice that the entire Process Design Procedure covers steps 6,8,9 and 11 of the Universal Design Procedure outlined in Fig. 2.1.

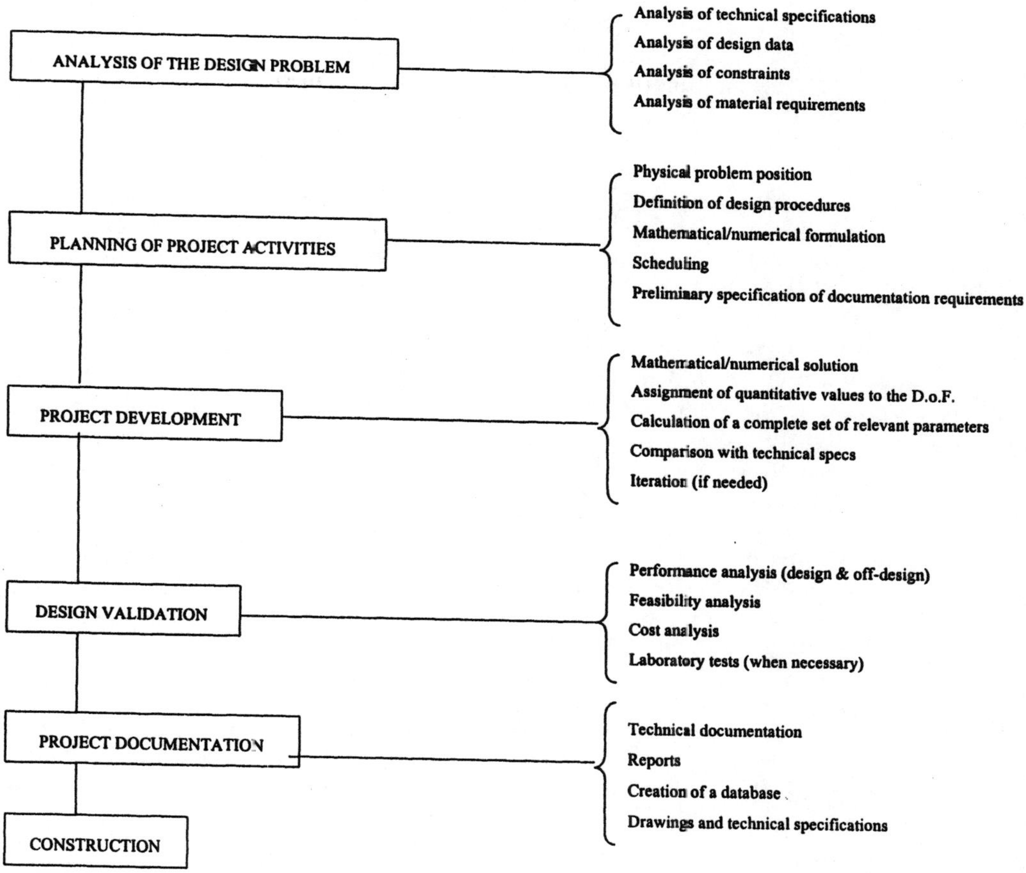

Figure 2-2. Process Design Procedure

2.3 - "DESIGN" AND "OPTIMIZATION". DIRECT AND INVERSE PROBLEMS.

The term *optimization* is one of the most misused concepts in the field of Engineering. When reading a technical report about some "optimization" performed on a process or a single component, it is often clear that, in the course of the "optimization" procedures, there was confusion about what was being optimized with respect to what, or what was to be kept constant. Most disappointing though is the fact that in the quest for an idealized optimum point, frequently configuration, operation and maintenance issues are entirely neglected or grossly underestimated, and the solution obtained via a purely mathematical procedure is presented as *the* solution to the given design-and-optimization problem. Thus, neighboring "quasi-optima" are disregarded, that in real engineering applications often represent the most convenient solution. Before considering a quite different approach that will be proposed here below, it is useful to review the terms in which a design-and-optimization problem is formulated.

Given an otherwise unspecified process (but the presentation applies as well to any Engineering System), its behavior will in general be described by a certain number *n* of *relevant variables* **x**. These variables span a domain, called the *design domain* **D**, that contains the range of operation of the process as a sub-domain $\mathbf{D_0}$. The *performance* of the process, however we wish to define it, is expressed by a usually much smaller set of *p performance parameters* **w** $(p << n)$ that individually and in some of their possible combinations depend on the *operational* point identified in the sub-domain $\mathbf{D_0}$ by *m* operational variables, and in **D** by the proper *(n-m)* design variables. The formulation of the problem also generally makes clear that there are a number *v* of constraints **c** that limit the range of variation of the **x**'s in **D** either directly or indirectly. An act of "Design" then consists in choosing the *n* design variables in such a way that all of the *v* constraints are abided by and all of the *p* performance parameters achieve satisfactory values. Notice that a design procedure also specifies a certain number *e* of correlations between the *n* variables, derived by phenomenological considerations or general balance laws, which limit the degrees of freedom (the breadth of the possible choices to be made by the designer) to $df = n - e$.

Let us define an *objective function* **f** as a real function of some or all of the **w**'s. The context of each specific problem clarifies whether this "global performance index" **f** is to be minimized or maximized to achieve more satisfactory results.[6] "Optimization" then consists in the following: *find the proper extremum of* **f**, *under the applicable set of constraints* **c**, *for the* **x** *belonging to the design domain* **D**.

It is important at this point to realize that, contrary to common opinion and in spite of the fact that the two topics are usually considered separately even in Engineering Design Manuals, *the activities commonly described as "Design" and "Optimization" are one and the same thing*: for, willingly or unwillingly, any person who generates a new concept or applies an old one to a different application, by "designing" some artifact or some process, is at the same time performing an "optimization", in that he tries implicitly to achieve the "best" solution under the technical/economical constraints he is faced with and according to his capabilities and knowledge. This does not mean that all the literature on optimization techniques is useless: it means rather that the concept of optimization must be somehow blended with that of design, to be able to treat both more effectively. It will become clearer in the following of this book that

[6] Notice that **f** may also explicitly contain some of the **x**'s

this new point of view is a cornerstone in the developing of Expert Design Assistants.

In practice, there are two forms in which "Design and Optimization" problems are encountered in Engineering: *direct* and *inverse* problems. A direct problem has the following formulation: given the process or system configuration, the proper description of all of its sub-processes (components) and the problem constraints, *find the operational point* **P(x)** *in the sub-domain* $\mathbf{D_0}$ *such that a "Design Performance Parameter", commonly called the objective function "f", attains its extremal value.* Examples of a direct problem are:

- "Seek the pressure ratio of a gas turbine cycle that maximizes process efficiency". Here, the objective function is the process efficiency, and the operational point P(x) can be chosen in a one-dimensional space (all the possible values of the process pressure ratio);
- "Find the minimum cost Heat Exchanger Network for a given set of hot and cold utilities". Here, the objective function is the overall (installation and operation) cost, and the operational point must be chosen in the finite set of the possible N configurations of the HEN;
- "Find the optimal load scheduling for a combined fossil-fueled powerplant operating in cogenerating mode". Here, the (likely!) objective function is the sum of the average costs of both the electrical and thermal energy produced by the plant over a year, and the operational point lies in a theoretically infinite, but in practice quite small, multi-dimensional space of feasible load curves.

An inverse problem has a different formulation: given the design objective and the process structure, *find the "optimal" design point* **P(x)** *in the domain* **D** *such that: a) the design objective is attained for the specified operational range* $\mathbf{D_0}$, *and b) the objective function* **f** *(which may be different from the one used in direct problems) attains an extremum.* Examples of an inverse problem are:

- "Seek the optimal design of a compressor which, for the given inlet conditions, delivers a specified mass flow rate at a specified pressure *and* has the highest possible efficiency". Here, the design objective consists of two data (flow rate and outlet pressure), and the objective function is the process efficiency;
- "Design a boiler given the required steam output and conditions, the inlet water conditions, the available fuel, and the specified level of allowable stack emissions". Here, the design objective is the required output, the specifications about the fuel and the water constitute a portion of the design domain, and the objective function, though not explicitly expressed, is clearly the minimum cost (including environmental taxes if applicable).

From a physical point of view, these two formulations are different. Yet, because they can be treated with a common mathematical approach, little attention is usually given (even by expert engineers) to the conceptual difference, and little emphasis is put in Engineering textbooks in clearly stating this difference. It is easy to put things in the proper perspective by considering the "optimization" of a compressor performed under the two approaches.

The direct formulation reads: "Given the geometry of the compressor, its performance curves, and the state of all thermodynamic quantities of interest, compute the operational point that corresponds to the maximum efficiency". It is clear that the search for the maximum efficiency is an easy task, which could actually be performed simply by inspecting the operational map of the compressor.

The inverse formulation reads: "For the prescribed values of the mass flow rate *m* and of the compression ratio β and for a given set of inlet conditions, find the (generally unique) set of geometric, cinematic and dynamic quantities that, under the surge and choke constraints, deliver the specified flow rate at the specified pressure with the highest possible efficiency". It is immediately clear that this task is much more difficult than the previous one, for one thing because *the set of the independent variables included in the optimization routine is much larger here than in the direct problem,* and furthermore because *the direct problem requires an unconstrained maximization (the only requirement is that the efficiency be at its maximum), whereas the inverse problem requires the optimization to be performed under a strong constraint* [45].

In engineering practice, we are often faced with an optimization problem of a different sort: "Given a certain input, find the *optimal process scheme to deliver a certain output (possibly varying in time) and specify correspondingly the optimal design point as well as the optimal operational schedule*[7]. This is a higher level task than the Design-and-Optimization problems discussed above. Now, we introduce an additional variable space **CS** (Configuration Space) whose extent may *a priori* be unknown and which contains *all of the technically possible embodiments of the process*. If we set equal to *r* the unknown number of these possible configurations, then the process synthesis and optimization problem can be thought of as an extremal problem whose solution is the locus of the *r* optima of each configuration. If the problem is non linear, there are very few examples of automatic approaches to its solution [1,5,8,12,20,31]. This consideration is reported here because, as discussed in the following Section, Artificial Intelligence techniques are well suited to tackle Synthesis-and-Optimization problems effectively.

2.4 - Computer-Aided Design.

With the advance of numerical techniques, virtually every procedure has been "computerized", so that engineers can avail themselves of a multitude of "computer tools" in the majority of their process design activities. There is however a frustrating lack of coordination among different procedures, so that the potential user, however experienced, is usually confronted with a very confusing situation when searching for a general simulation tool. There are in

[7] The problem can be further complicated by the requirement of finding the optimal process scheme for a certain set of *operational* points: but this will be not considered here.

fact many codes which perform nominally the same task, giving at times substantially different results when applied to the very same problem [65]; in addition, each code uses its own set of property tables, its own input/output format, etc. Finally, with very few exceptions, these codes are not mutually compatible, i.e., the output from any of them cannot be used as the input to any of the other, not only because of the respective formats, but rather because the quantities taken to represent a certain physical process are not the same in different codes.

It is obviously in the best interest of Design- and Process engineers that this conceptual gap be bridged: this can be achieved by detecting all similarities in known design procedures, constructing a suitable knowledge base, and implementing a general modular procedure that can assist the engineer in the largest possible number of process design calculations. The actual possibility of performing this task is here demonstrated by means of an example [65]: we begin by describing a simple process calculation as it would be performed using a hand calculator and property tables, and then proceed from there toward a fully interactive and entirely modular process simulator. A simulator is a code (a "package" in commercial computer jargon) that possesses a built-in capability of executing all the calculation tasks an engineer would perform, once the structure of the process (its "configuration") is known. This is, in practice, the state-of-the-art of computer-aided design for what thermal systems are concerned. But, the real possibility exist of going one (conceptual) step further, and implement a code capable of choosing the structure of a process best suited for a given application: this is the realm of "Process Synthesizers", and, while it forms the major portion of this book, nevertheless is a topic technically and logically in its infancy. This is so true, that even the name "Process Synthesizer" is proposed here for the first time. These codes rely heavily on AI techniques, and consist mainly of qualitative "computations", but can be interfaced with strictly numerical codes, to produce quantitative data in a format directly amenable to design applications. The overall path is clear: we are trying to extract from the various engineering activities all the essential knowledge that pertains to engineers, with the final goal of transferring this body of knowledge, in some form suitable to machine communication, to a "Universal Process Synthesizer/Simulator", that can then be applied with a high degree of confidence to a wide variety of Process Engineering problems.

2.4.1 - Unit- and Process Simulators

Consider a relatively simple engineering problem: the design of a combined heat and power plant, featuring a gas turbine/generator set to produce the required electrical power P_{el} and a waste heat recovery boiler -fed by the gas turbine exhaust gas- to produce the required heat rate Q[8]. The problem is properly formulated: both P_{el} and Q are specified, and the additional design constraints are complete and congruent. The fuel is gas turbine oil with a known lower heating value of LHV. The maximum permissible pressure ratio and maximum safe turbine inlet temperature have been chosen. The design values for the air and water physical properties have been specified, and the emission limits properly set.

Suppose we have no other computational means than a hand calculator and a (possibly graphical) set of air and combustion gas properties (like the "Air Tables" reported in most textbooks of Thermodynamics, for example. But see also [54]). The calculation proceeds as follows (the suffixes refer to Fig. 2.3):

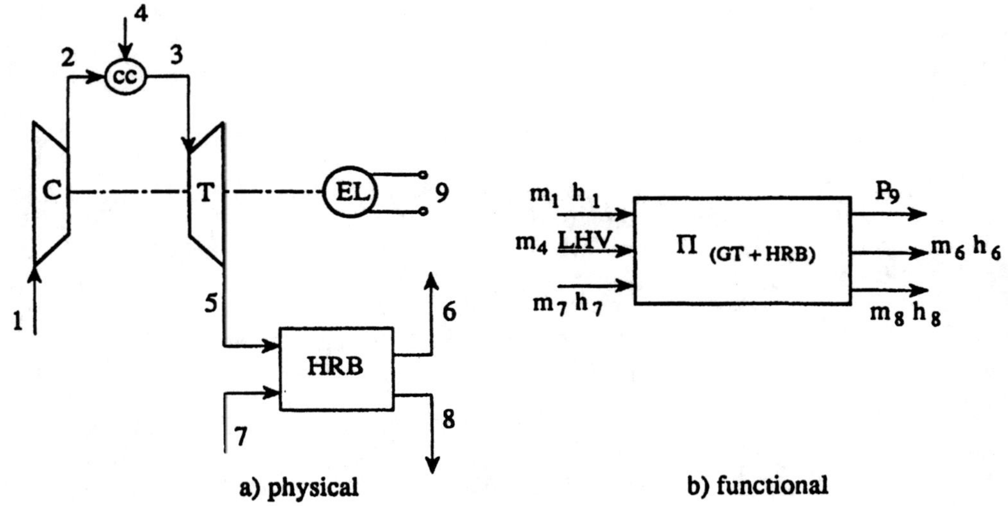

Figure 2-3. Combined heat and power plant

[8] This example is taken from [65], but it was based, in its original version, on a general design problem proposed in [44]

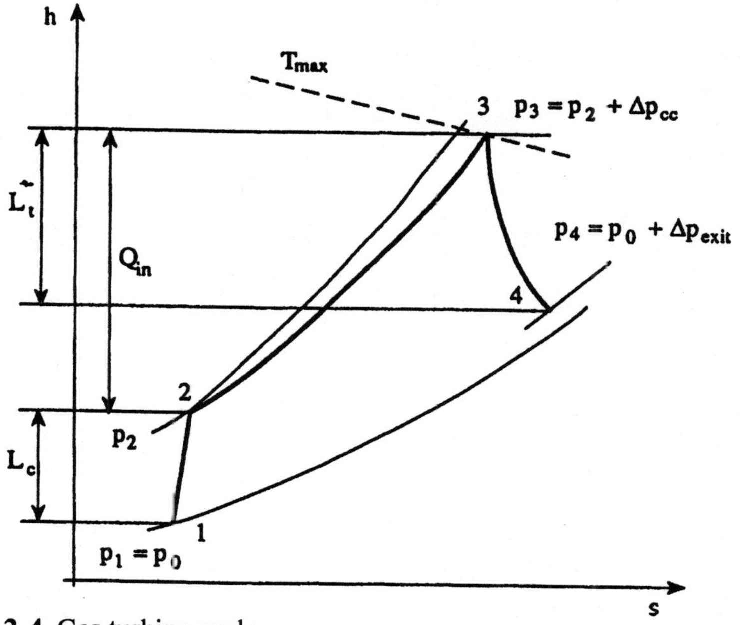

Figure 2-4. Gas turbine cycle

A - Compute the gas turbine cycle (Fig. 2.4)

step 1) Find T_2, the temperature of the compressed air entering the combustion chamber:

$$T_2 = f_1 (T_1, p_2, p_1, k_{air}, \eta_{c,is}) \qquad 1)$$

step 2) Find p_3, the pressure of the combustion gas exiting the combustion chamber:

$$p_3 = p_2 - \Delta p_{2\text{-}3} \qquad 2)$$

step 3) Specify T_3, the maximum allowable gas temperature at turbine inlet:

$$T_3 = T_{max} \qquad 3)$$

step 4) Compute the fuel-to-air-ratio:

$$\alpha = \frac{m_4}{m_2} = f_2(T_2, T_3, c_{p,2-3}, \eta_{cc}) \qquad 4)$$

step 5) Compute T_5, the gas turbine exit temperature:

$$T_5 = f_3\,(T_3, p_3, p_5, k_3, \eta_{t,is}) \qquad 5)$$

step 6) Compute the specific turbine work output:

$$w_t = (1+\alpha)\,(h_3 - h_5) \approx (1+\alpha)\,(c_{p,3}\,T_3 - c_{p,5}\,T_5) \qquad 6)$$

step 7) Compute the specific compressor work requirements:

$$w_c = h_2 - h_1 \approx c_{p,2}\,T_2 - c_{p,1}\,T_1 \qquad 7)$$

step 8) Compute the net specific electrical output:

$$w_{el} = w_9 = (w_t - w_c)\,\eta_m\,\eta_{el} \qquad 8)$$

step 9) Compute the required mass flow rate of air:

$$m_1 = P/w_{el} \qquad 9)$$

B- Compute the steam generating process

step 1) Compute the throttle steam temperature (ΔT_{pinch} is specified):

$$T_8 = T_5 - \Delta T_{pinch} \qquad 10)$$

step 2) Compute the enthalpy of the water at boiler inlet:

$$h_7 = f_4(T_7, p_7) \qquad 11)$$

step 3) Compute the enthalpy of the steam at condenser inlet:

$$h_8 = f_5(T_8, p_8, x_8) \qquad 12)$$

step 4) Compute the steam mass flow rate:

$$m_8 = Q/(h_8 - h_7) \qquad 13)$$

step 5) Compute the gas temperature at stack inlet:

$$T_6 = \frac{1}{c_{p,6}}\left[c_{p,5}T_5 - \frac{Q}{\eta_{hrb}(1+\alpha)m_1}\right] \qquad 14)$$

In the above equations, f_1 ... f_5 are thermodynamic relations involving the isentropic compressor and turbine efficiencies, the energy balance for the combustion chamber, the definition of enthalpy, etc.: they are well-known expressions and can be derived in closed form. However, if graphical property data are available for air and water, then equations 1) and 5), as well as 6) and 7), can be replaced by an easy geometric construction (Fig. 2.4), and equations 11) and 12) can be entirely dispensed with, because enthalpies are directly read off the Mollier diagram.

With no lack of generality, the isentropic efficiencies, the ΔT's and Δp's can be thought of as "design constraints" and assumed as known. Notice that there are some intermediate checks that have to be performed: for instance, if $w_c > w_t$, it will be necessary to iterate between steps 1) and 7) with different values of p_2. Or, if T_6 turns out to be lower than the minimum acceptable stack gas temperature, it will be necessary to lower T_8 or provide the heat recovery boiler with an afterburner.

The foregoing calculations are simple, and, with a minimum of engineering experience on the part of the designer, always result in a well-defined process scheme. This type of task can become very tedious though, especially if different sets of design values have to be considered. Furthermore, hand-based calculations are very much prone to human error ("miscalculations"). Note, in particular, that equations 1), 4) to 7), 11), 12), 14) require property values that have to be extracted from the air- and steam tables: both the numerical and graphical interpolations can be sources of errors. Anyhow, the above calculation represents a proper *process calculation procedure*, and is an example of what we want to replace with an automatic, computer-assisted method.

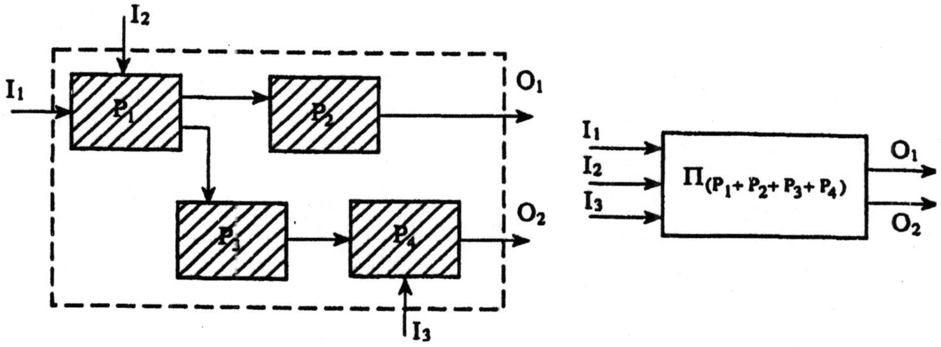

Figure 2-5. Definition of the Process Transfer Operator Π

2.4.1.1 - The notion of "Process Simulation"

To limit the scope of the investigation and to make the goals more precise, let us give a general definition of the engineering activity that goes under the name of *Process Simulation* (of which an elementary example was given above). Denote by **P** an otherwise unspecified physical process and assume that the task is to calculate all the relevant quantities that identify **P** *for all practical engineering purposes*. This means that we will in general be concerned with problems that can be treated using the continuum hypothesis, the principles of Engineering Thermodynamics, etc.. Processes that require a relativistic description or which by their nature defy the notion of continuum, etc. are not explicitly considered. However, such exceptions can be easily treated by relaxing some of the basic constraints.

Whatever the complexity of a process, it is always possible to break it down into sub-processes that are simpler, either because they are composed of a lower number of "elementary" thermodynamic transformations or because they refer to a smaller number of participating media. By repeating this splitting procedure as necessary, it is possible to reach a point at which the physical description of the single, complex process **P** can be shown to be equivalent to the description of several simpler interconnected sub-processes $\mathbf{P}_1 \ldots \mathbf{P}_n$. With no lack of generality, we will assume here that this decomposition has already been performed; and that, once the sub-processes have been simulated, it will be possible to re-assemble these $\mathbf{P}_i$'s into the original process **P**.

As shown in Fig. 2.5, the description of **P** can be in principle represented by an *operator* Π , which *acts on the inputs to produce the outputs*. Thus, "calculating" **P** means finding a suitable formal expression for Π (called the *transfer function* of **P**), applying it to the inputs $\mathbf{I}_j$ (j=1...f) and deriving numerical computable formulae for the outputs $\mathbf{O}_k$ (k=1...p), where **O** and **I** denote the output- and input vector respectively:

$$\mathbf{O} = \Pi \langle \mathbf{I} \rangle \tag{15}$$

In the most general case, Π is a nonlinear matrix operator linking *all* of the outputs with *all* of the inputs. No limit has been imposed on the form of operator Π: it can be algebraic, differential, integral-differential, steady or unsteady, etc. The form that Π takes will of course determine the number and type of necessary boundary (and/or initial) conditions, i.e. a certain number of design data which quantitatively define some of the inputs or outputs.

On the basis of these considerations, we are now in a position to give an exact description of Process Simulation:

Given a certain well-defined physical process, we aim to
i) devise a suitable representation for it in terms of its sub-processes, if necessary,
ii) find a formal mathematical representation for its transfer function Π,
iii) acquire all necessary boundary and initial conditions,
iv) solve the system 15) for each sub-process, and
v) assemble the partial solutions into the global solution

In the remaining of this Section, each one of these process simulation steps is analyzed in some detail, with the objective of extracting a sufficient number of general principles that can be implemented in *automatic* numerical procedures, rather than by hand-calculations such as the one just described.

2.4.1.2.-Automatic calculation of material properties

The first step to make the process simulation less dependent on human resources is that of developing "material properties routines" that provide the thermodynamic quantities of interest (enthalpy, entropy, saturation values, steam quality in the wet region, specific heats, etc.). These calculations are carried out on the basis of interpolating formulae that can be *specific* for the particular fluid under consideration, or of *general validity* for a certain set of fluids.

Both types of routines are presently available, and both can be developed in such a way as to yield a high degree of accuracy while still retaining a relatively high degree of efficiency. Since deficient property routines can be very detrimental for the overall accuracy of the calculations, it is advisable to use only well-tested and well-documented packages (like [59], for example). These routines can be used either in "manual" mode, i.e. by plugging into them the values of the known parameters and deriving the outcome, or in "automatic" mode, i.e. by using them on some kind of calculator (for example, one of the hand-held programmables). If a set of these routines is available, equations 1) and 5), and consequently 6) and 7), can be directly computed in closed form, and the same applies to equations 11) and 12). The advantages in terms of the repetitivity and of the accuracy of the solution are apparent: still, the entire procedure is essentially based on hand-calculations that are time intensive and prone to error.

2.4.1.3- Automatic calculation of thermodynamic relations

In our quest for a fully automatic procedure for the calculation of the combined steam/gas cycle, let us now turn to the possibility of implementing a numerical procedure for the calculation of the thermodynamic relations, equations 1) to 14). Here, we are immediately faced with a fundamental problem common to *all* numerical process calculators: it is not difficult to solve equations 1)...14) sequentially (i.e. in the very same order in which they would be solved by a manual procedure) *but it is difficult to formulate a general procedure to solve the system of equations 15) by direct inversion of the matrix Π without "knowing" in advance the structure of the process.* Let us explicitly compute Π for the present application. Referring again to Fig. 2.3, which shows clearly that m_1h_1, m_4 LHV and m_7h_7 are the *inputs* and $P_{el} = P_9$, $Q_{stack} = Q_6$ and $Q_{st} = Q_8$ are the *outputs*, we can write equation 15) in its expanded form, and obtain (with $\alpha = m_4/m_1$, $\kappa = m_7/m_1$):

$$O_1 = P_{el} = [(1+\alpha)(h_3 - h_5) - (h_2 - h_1)]\eta_{el}\, m_1$$
$$O_2 = Q_{stack} = (1+\alpha)\, m_1\, h_6 \qquad 16)$$
$$O_3 = Q_{st} = m_1 \kappa(h_8 - h_7)$$

Which has the general structure indicated by equation 15):

$$O_1 = \Pi_{11} I_1 + \Pi_{12} I_2 + \Pi_{13} I_3$$
$$O_2 = \Pi_{21} I_1 + \Pi_{22} I_2 + \Pi_{23} I_3 \qquad 17)$$
$$O_3 = \Pi_{31} I_1 + \Pi_{32} I_2 + \Pi_{33} I_3$$

and corresponds to the following form for Π (which is, as stated above, the transfer function of the process):

$$\Pi = \begin{vmatrix} \pm\dfrac{(h_2 - h_1)\eta_{el}}{h_1} & \dfrac{(1+\alpha)(h_3 - h_5)\eta_{el}}{\alpha LHV} & 0 \\ 0 & \dfrac{(1+\alpha)h_6}{\alpha LHV} & 0 \\ 0 & \dfrac{(1+\alpha)(h_6 - h_5)}{LHV}\eta_{hrb} & 1 \end{vmatrix} \qquad 18)$$

Notice that with the same approach it is possible to compute the mass and energy balances of a component or of a cycle. In fact, equations 16) are just another way of writing the energy balance. We are making use here of the well-known black-box approach: the description of the generic process **P** (the cogenerating process) in terms of its sub-processes (compression, water heating-boiling-superheating, Section 2.4.1.1 is exactly the same approach adopted in the input/output (or black-box) process representation.

Since h_2 $(h_1,\beta,k_{12},\eta_c)$ and all other thermodynamic relations cannot be computed unless they are expressed by explicit functions, one must augment the Π defined by the matrix 18) with a sufficient number of auxiliary equations of the type:

$$h_2 = h_2(p_2,T_2)$$
$$h_5 = h_5(p_5,T_5) \qquad 19)$$
etc.

It can be shown [58] that a system in the form of equations 17) can always be explicitly written for any process, and that it can always be derived using mass- and energy balance equations in symbolic form (i.e. written without using any numeric, just in the same way in which they would be written on a process sheet). These auxiliary equations make the calculation procedure *highly specialized* for the cogenerating gas turbine process, so that *the same procedure cannot be employed to compute any other process (and not even similar processes with slightly different configurations, such as a regenerated or intercooled gas turbine cycle)*. That is, there *are as many independently derived* **Π 's** *as the number of individual cycles or processes we wish to simulate; and these Π 's are in general different from each other*. This is actually the present situation: there are many codes for the process calculation of different processes or cycles, but they are very much application-oriented and lack generality. [9]

2.4.1.4- Towards Modular Simulators

The conclusion that can be drawn from the foregoing considerations is that *to develop a general process simulator, it is necessary to impart to the code a flexible enough structure to be able to automatically handle physically different processes*. This is where the concept of a modular process simulator comes into the picture. If a number of elementary sub-processes $\mathbf{P}_i$ can be identified that can be used as building blocks for *any* process **P** of interest, a general process simulator can be constructed in principle by the following procedure:

i- Create a *passive database* of elementary modules, each one of which simulates a process $\mathbf{P}_i$
ii- Construct a user-machine (software) interface capable of reproducing the physical interconnections among the various $\mathbf{P}_i$'s
iii- Solve each $\mathbf{P}_i$ in turn under the proper assembly constraints, i.e. take into proper consideration the interactions of $\mathbf{P}_i$ with all other $\mathbf{P}_j$
iv- Reassemble the solution under the global structure of **P**.

Such a process simulator will have by necessity a highly modular structure, consisting of (1) a *core* that handles the I/O and implements the general solution strategy, (2) a series of subroutines which are called in turn to solve a particular sub-process, and (3) a series of property-tables in a database. The following Sections describe the general features of the various portions of a possible code of this kind.

2.4.1.5- Translation of the process structure into machine-readable input

The physical structure of a process can be represented by a *process diagram* (Fig. 2.6), which displays the general layout of the system that implements the process. This diagram identifies the components and the fluxes of matter and

[9] Most of the present "*Modular Simulation Packages*", like ASPEN^{tm}, GATECYCLE^{tm}, etc., are only modular in the user-interface, and not really so in their logical construction.

energy that connect the components to each other, describes the exchanges of the system with its immediate surroundings, and contains essential quantitative information about the operating parameters of the process (pressures, mass and energy fluxes, enthalpies, etc.). A numerical simulator cannot "read" the process diagram as a human does. It is therefore necessary to devise a "computer-readable" form of this diagram which ought to be easily accessible by the code, but not too removed from the intuitive mathematical comprehension of the user. The most suitable form is that of a matrix **IM**, defined as follows [5,36]:

$IM_{i,j}$ is a character string that contains the *name of the component*
$IM_{i,1}$ is a character string that contains the *identifier of the flux*
$IM_{i,j}$ is = 1 if flux i *enters* component j
$IM_{i,j}$ is = -1 if flux i *exits* component j
$IM_{i,j}$ is = 0 otherwise

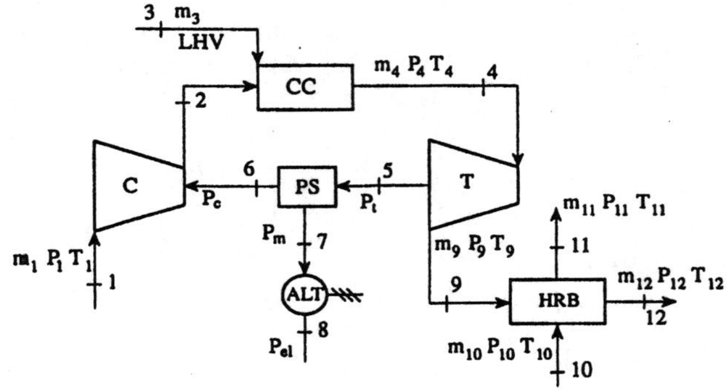

(a) Process scheme

	C	CC	T	PS	ALT	HRB
1-	1	0	0	0	0	0
2-	-1	1	0	0	0	0
3-	0	1	0	0	0	0
4-	0	-1	1	0	0	0
5-	0	0	-1	1	0	0
6-	1	0	0	-1	0	0
7-	0	0	0	-1	1	0
8-	0	0	0	0	-1	0
9-	0	0	-1	0	0	1
10-	0	0	0	0	0	1
11-	0	0	0	0	0	-1
12-	0	0	0	0	0	-1

(b) Interconnection matrix

Figure 2-6. Process diagram for the combined powerplant of Fig. 2-3

The matrix **IM** is the representation of *the connectivity graph* of the system (augmented by the inessential character strings) and is thus properly called the *interconnection (or structural) matrix*. This matrix is easy to construct once the structure of the system has been identified. An example of the "translation" of a process diagram into the corresponding **IM** is shown in Fig. 2.6 for the combined-cycle plant of Fig. 2.3. Notice the different numbering of the flows in these two figures, due to the slightly different structure of the logical process scheme [65][10].

Besides constituting a synthetic way of describing the process structure, **IM** has another very important characteristic: *it allows a very efficient formal representation of the mass and energy balances of a system.* Consider again the process scheme of Fig. 2.6 and denote by m_1, m_2, etc. the mass flow rates in the sections 1,2... respectively. It will also be possible to define a "mass flow vector", **m**, whose entries are the various mass flow rates, with the additional stipulation that the fluxes which do not transport any mass (typically, as in this case, the "power" fluxes) are assigned a value 0. The product:

	comp	cc	gt	ps	alt	hrb	/ / /
	1	0	0	0	0	0	m_1
	-1	1	0	0	0	0	m_2
	0	1	0	0	0	0	m_3
	0	-1	1	0	0	0	m_4
	0	0	-1	0	0	1	m_5
IM*m =	1	0	-1	1	0	0	0
	0	0	0	-1	0	0	0
	0	0	0	-1	1	0	0
	0	0	0	0	-1	0	m_9
	0	0	0	0	0	-1	m_{10}
	0	0	0	0	0	1	m_{11}
	0	0	0	0	0	-1	m_{12}

is equal to the *mass balance vector*:

comp	$m_1 - m_2 = 0$	
cc	$m_2 + m_3 = m_4$	(if component is a *mixer*, a "split" mass balance is convenient)

[10] The physical representation assumes that the compressor absorbs from the gas turbine just the amount of power it needs, and that the remaining portion is delivered to the electrical generator. It would be very difficult to impart this notion to a "Process Simulator": therefore, if a situation arises in which a certain amount of power must be apportioned to two different devices, an additional component (a "power splitter", PS) is inserted. For similar reasons, the use of "Flow Splitters" is also quite common in Process Simulators. The opposite action (merging of two streams or joining of two shafts) is performed by Power Mergers and Flow Mixers

gt	$m_4 - m_5 = 0$	
ps	0	
alt	0	
hrb	$m_5 - m_{10} = 0$; $m_{11} - m_{12} = 0$	(mass balance is written independently for unmixed flows)

A similar formalism applies to the energy balance as well: denote by h_1, h_2, etc. the specific (stagnation) enthalpies in the sections 1,2... respectively, and define an "energy vector", **mh**, whose entries are the various enthalpies multiplied by the relevant mass flow rates. For electrical and mechanical power fluxes, the values (P_9 etc.) are directly inserted; for fuels (here, flux number 4), the entry is computed as the sum of the stagnation enthalpy and of the LHV. The product:

IM*mh =

comp	cc	gt	ps	alt	hrb	/ / /
1	0	0	0	0	0	m_1h_1
-1	1	0	0	0	0	m_2h_2
0	1	0	0	0	0	m_3h_3
0	-1	1	0	0	0	m_4h_4
0	0	-1	0	0	1	m_5h_5
1	0	-1	1	0	0	P_6
0	0	0	-1	0	0	P_7
0	0	0	-1	1	0	P_8
0	0	0	0	-1	0	P_9
0	0	0	0	0	-1	$m_{10}h_{10}$
0	0	0	0	0	1	$m_{11}h_{11}$
0	0	0	0	0	-1	$m_{12}h_{12}$

is equal to the energy *balance vector*:

comp	$m_1h_1 - m_2h_2 = P_7$	(energy fluxes of different type are on different sides of the = sign)
cc	$m_2h_2 + m_3h_3 = m_4h_4$	(reaction products and components are on different sides of the = sign)
turb	$m_4h_4 - m_5h_5 = P_6$	(energy fluxes of different type are on different sides of the = sign)
ps	$P_6 - P_7 + P_8 = 0$	
alt	$P_8 - P_9 = 0$	
hrb	$m_5h_5 - m_{10}h_{10} = m_{12}h_{12} - m_{11}h_{11}$	(energy fluxes of unmixed flows are on different sides of the = sign)

This fact has important consequences in the programming, because it makes possible the use of very compact and modular system assembly methods.

CAMEL: modular elemental plant calculation version 2.1
Input for the calculations; matrix IPM

1 -	447.00	100.00	300.00	.21	AIR
2 -	.00	.00	.00	.00	FTU
3 -	.00	41840.00	.00	14.20	FTU
4 -	.00	832.00	1370.00	.00	FTU
5 -	.00	.00	.00	.00	FTU
6 -	.00	.00	.00	.00	FTU
7 -	.00	.00	.00	.00	FTU
8 -	.00	.00	.00	.00	FTU
9 -	.00	.00	.00	.00	FTU
10 -	60.00	120.00	760.00	1.00	SHS
11 -	.00	.00	433.00	.00	WAT
12 -	.00	5290.00	.00	.00	FTU

CAMEL: modular elemental plant calculation version 2.1
Output for the calculations; matrix PM

1 -	447.00	100.00	300.00	.21	AIR
2 -	447.00	857.73	583.08	.21	AIR
3 -	11.65	41840.00	38.34	14.20	FUE
4 -	458.66	832.00	1370.00	.13	GAS
5 -	253214.09	.00	.00	.00	POW
6 -	131838.02	.00	.00	.00	POW
7 -	121376.06	.00	.00	.00	POW
8 -	115307.26	.00	.00	.00	POW
9 -	458.66	120.00	791.15	.13	GAS
10 -	60.00	5131.00	760.00	1.00	SHS
11 -	60.00	5290.00	433.00	.00	WAT
12 -	458.66	116.00	458.67	.13	GAS

Figure 2-7. IPM and OPM for a combined powerplant

Since the balance equations constitute a nonlinear system, an iterative solution procedure is used. To begin with, an initial "state" of the system must be specified. This initial state does not need to be consistent (in the sense that the values of the assigned parameters may also *not* satisfy the balances) and usually contains design data and some additional information about known or guessed values for the process parameters. To impart these data to the code, it is necessary to assign *in the correct sequence* the various parameters (pressure, temperature, mass flow rate, composition, etc.) to the proper sections of the process diagram. This initialization can be done very efficiently by means of a series of specific subroutines that contain in a structured sequence relevant information about each component, its location in the system, and the working fluids through it. This "structure" is then used by the code to properly identify the data contained in two structured data-tables, which we will denote as the *input plant matrix*, **IPM** and the *output plant matrix*, **OPM**. It would be advisable that both the **IM** and the **IPM** be constructed with the aid of a specific utility, to facilitate the user-friendliness of the simulator; the **OPM** must be automatically generated by the code. For a detailed description of these matrices, see [9]. The **IPM** and **OPM** of the combined-cycle case study considered here are reported in Fig. 2.7.

For unsteady state or transient simulations, the matrices **IPM** and **OPM** will in general be time-dependent so that there will be an **IPM**(0) and an **OPM**(0), which will represent the *initial state* of the system just before the transient begins. If there are *n* time steps in the integrating procedure, there will be a series of *n* matrices **IPM**(t_1)...**IPM**(t_n), **OPM**(t_1) ...**OPM**(t_n), which will contain all the process information needed at each time step. It is important to remark that this transient procedure rests on the assumption that *every unsteady process can be represented via a quasi stationary approximation.* This point will be discussed again in Section 2.4.1.8 here below.

2.4.1.6- Solution of the global system of balance equations

The nonlinear system of equations resulting from the application of the mass and energy balances to every component of the process does not constitute, except in trivial cases, a solvable system because its coefficient matrix is not square [58]. However, it is *always* possible to augment this coefficient matrix by the use of property relations (see Section 2.4.1.3 above) and by the imposition of the design specifications, which act as *constraints* on some of the degrees of freedom of the process. The final system is in the form

$$\mathbf{CM} * \mathbf{x} = \mathbf{d} \qquad 20)$$

where **CM** is the augmented coefficient matrix, **x** is the vector of unknowns and **d** is a vector containing the design specifications and the constraints. The rank *r* of **CM** is considerably *higher* than the number of components, because there is generally more than one equation per component, and because additional equations have been added to make the system solvable (i.e. **CM** invertible).

The system of equations 20) has two types of nonlinearities: one is explicit and is due to the fact that some of the equations (e.g., the isentropic relations) are nonlinear in a formal way, i.e., they contain one or more variables raised to a rational power or under logarithm or exponential; and one is implicit, because the property relations [e.g., $c_p = f(T)$] are nonlinear in a physical sense, i.e., the variable of interest depends from some of the thermodynamic variables with a law of dependence which is nonlinear. Therefore, a stable and robust iterative procedure must be employed to solve the system. There are several possible schemes available in the literature [9,53].

2.4.1.7- The Selectively Segregated approach for steady-state simulations

Non-linear systems of equations can be solved by direct inversion of the coefficient matrix: the non-linearity requires the use of iterative procedures. In Process Simulation, a different solution method, called "Component Oriented Solver", is often employed: it consists of solving in sequence only the sub-matrix pertinent to each single component of the process. The advantages claimed by the supporters of this method are twofold: first, the rank of each submatrix being substantially lower than that of **CM**, it may well be that the global solution time is reduced; second, it is easier to intervene at "component" level to fix numerical convergence or stability problems. These "advantages" are more illusory than real though, because thanks to the continuously increasing clock performance of modern computers, the difference in real CPU time between a Component Oriented Solver and a global solver (also called "Equation Oriented Solver") is reduced to seconds; furthermore, it is debatable whether it is better to force each component in turn to convergence, and then iterate anyhow on the entire process (because it is highly unlikely that at least one of the assumed inputs in the *j-th* component be not modified by the results of the calculation performed on the *(j-1)th* or *(j-2)th* component), or to accept a more CPU-intensive procedure in which the iterative process is carried out automatically on the entire set of equations. Sometimes, it is the commercial format of the individual packages that dictates the solution strategy: so, "modular" packages like GateCycle or SimuLink possess a component-oriented solver, while complex packages like ASPEN+ or ChemCad adopt an equation-oriented procedure.

A different solution strategy for the system of equations 20) is though possible: one could *solve one sub-process after the other in an order that is not pre-determined, but rather dictated by the developing solution itself.* The algorithm has been called the Selective Segregated Solver [9] ("SSS" in the following), and is described below. The attribute *segregated* is borrowed from the field of numerical fluid dynamics, and serves as a reminder of the fact that it solves only *some* of the equations at each step, and then uses the partial solutions to sequentially update the set of "known" variables.

After reading the initial data from the matrix **IPM**, the SSS scans the components sequentially, starting from the first column of the matrix **IM**, until it finds one for which there are as many unknown variables as there are independent equations. Let us assume that this j-th component is a compressor. It will be identified by 7 variables (m_1,p_1,T_1,m_2, p_2,T_2,w_c), four of which (m_1,p_1,T_1,p_2) are given in **IPM**. There are three equations available (the mass balance, the energy balance and the polytropic compression equation). SSS will start an iterative procedure and drive to convergence, thus completely specifying the operating parameter values of the compressor. SSS will then go on to the *(j+1)*-th component (combustion chamber) and determine that it has 8 independent variables and only 6 available equations. Accordingly, SSS will *skip* this component and proceed to the *(j+2)*-th one, and so on, until one of two conclusions is reached: either there are no more components to scan, and the **OPM** has still some undefined entries; or all the variables of the problem have been attributed a value. In the second case, the external iteration (extended to the entire system) is obviously completed; in the first case, it is not possible to solve the problem with the given data. More input data or more constraints are needed, and the user is correspondingly prompted.

When the *n*-th iteration has been completed, the resulting values are combined to those of the *(n-1)*-th step, to obtain the *corrected initial values for the (n+1)-th iteration*. For the generic variable ξ, the following formula applies:

$$\xi^* = \theta\, \xi_{n-1} + (1 - \theta)\, \xi_n \qquad 21)$$

where ξ^* is the initial value for the next iteration, and θ is an *underrelaxation factor*, whose function is to increase the stability of the numerical procedure by "damping" possible nonlinear oscillations in the solution. The range for θ is 0.1 to 0.9, with 0.5 a commonly employed value. The lower the underrelaxation factor, the faster the solution will converge; but the less stable it will be. A flow chart of the steady-state procedure is shown in Fig. 2.8 [65]. It is important to remark that the segregated solution strategy represents an improvement over the conventional equation-oriented solvers, that are obscure to the user and that require computer resources exponentially growing with the number of components to solve for. At the same time, the SSS is much faster than the existing component-oriented solvers, that employ a strictly sequential approach and do not "skip" components.

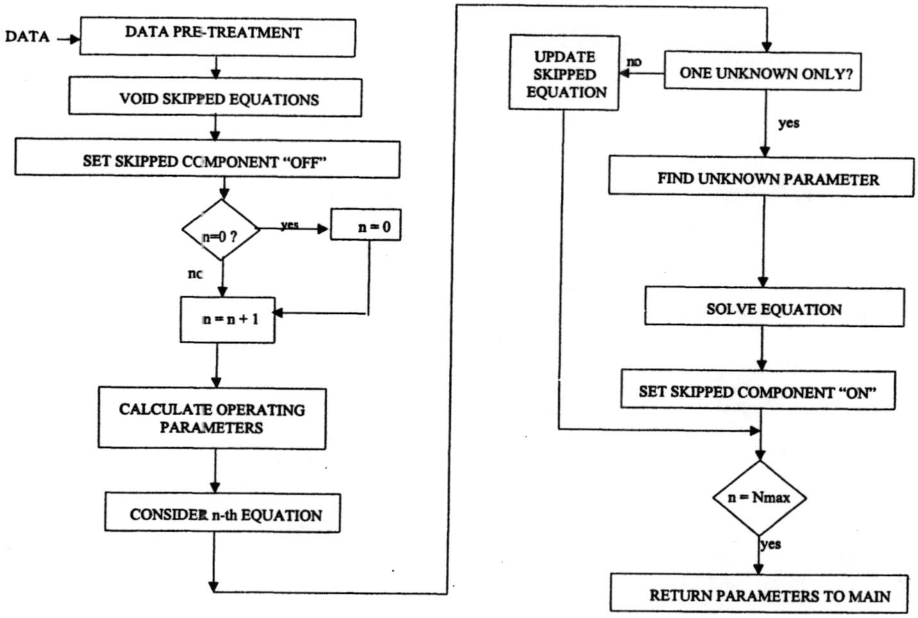

Figure 2-8. Flow chart of the steady-state Selective Segregate Solver

2.4.1.8- Unsteady simulations: a time-marching algorithm

The numerical simulation of unsteady processes is a very complex task, because it requires knowledge of a series of *instantaneous* data about the process:

I. the variation of ALL of the inputs that contribute to the transient operation must be known at each time step;

II. the *transfer matrix* of each component of the process must also be known; if $\mathbf{O}_{k,j}$ is the *k*-th output (k = 1...N) of the *j*-th component and $\mathbf{I}_{i,j}$ the *i*-th input (i = 1...M), the transfer matrix **F** is a (NxM) table which contains the single (now, time-dependent!) transfer functions Π of the individual components and takes the general form:

$$\Pi_j = \begin{vmatrix} \Pi_{1,1} & \Pi_{1,2} & \ldots & \Pi_{1,M} \\ \Pi_{2,1} & \Pi_{2,1} & \ldots & \Pi_{2,1} \\ \ldots & \ldots & \ldots & \ldots \\ \Pi_{N-1,1} & \Pi_{N-1,1} & \ldots & \Pi_{N-1,1} \\ \Pi_{N,1} & \Pi_{N,1} & \ldots & \Pi_{N,1} \end{vmatrix} \qquad 22)$$

where the individual entries are the coefficients of the general expression of the output-input relation:

$$O_{k,j}(t)=\sum_{i=1}^{M}\left\{\Pi_{i,j}(t)I_{i,j}(t)\right\} \qquad (k=1...N) \tag{23)}$$

and the Π-coefficients can be non-linear functions of several $\mathbf{I}_{i,j}$. Notice that the system 22) can be directly related to the matrix **C M** defined in Section 2.4.1.6.

III. the time response of EACH component of the system must be known, so that the proper time-shift can be inserted in the relevant Π-coefficients.

Except in relatively simple cases, a transient simulation can become a very cumbersome task. Furthermore, a procedure capable of performing *one specific* simulation would easily lack generality, i.e. it would be unsuitable for different processes or even for the same process undergoing a different transient. A substantially simplified, albeit less exact, approach is that of postulating that:

- the system proceeds from its *initial* to its *final* state through a succession of *local quasi-equilibrium states.* This means that, though the transfer matrix can be different at different time steps [i.e., $\Pi(t) \neq \Pi(t+\delta t)$], there is no *explicit* time-dependence in the Π-coefficients.
- it is always possible to choose a time step δt small enough that the response of every component is either instantaneous or displays a fixed delay in the vicinity of that δt: this means that the response can be computed in terms of ΔΠ instead of ∂Π/∂t.
- The transfer matrix Π_j does not depend on the transfer matrix Π_k of other components possibly connected to component "j".

On the basis of the above, rather stringent assumptions, a time marching procedure can be developed as follows: given that the entire state of the process is known at time $t_0 = 0$, assume for simplicity that the transient-driving force is a variation in only *one* of the process inputs. For each component, it is possible to compute the N outputs by either equations 22) or 23). It is convenient to begin the time-marching *from the component "J" which is the recipient of the time-dependent input*. Provided the δt has been chosen properly, the time delay introduced by component J will either be zero or a finite (small) value τ_J, and a poynomial response can be assumed, i.e.

$$O_J(t_0+\delta t)=f\left\langle \frac{\tau_J-\delta t}{\tau_J}, O_J(t_0)\right\rangle \tag{24)}$$

Once $O_J(t_o+\delta t)$ has been computed, the algorithm will proceed as if the simulation was at steady-state, and compute all the parameters that characterize the state of the system at the end of the first time step. Then t is set to $t_o + 2\ \delta t$, and the process is repeated.

When using the above procedure, the characteristic response of a component at a given time can be derived directly from its "performance chart", which is readily available for most components. Care is needed in regions prone to unsteady operation (e.g., near the choke or surge lines of a compressor) or where the characteristic response curve is steep. In both cases, it is possible that the search for a stable solution fails, because

1) near unsteady operation regions, it is possible that one of the variations of an input (for example, a reduction of the inlet mass flow rate in a centrifugal compressor) drives the component into the unsteady region, where, of course, the transfer matrix takes a different (and often unpredictable) form;
2) if the characteristic curve is steep (like for instance in the portion of the operational map of a gas turbine in the proximity of the choking limit), a small variation of one of the inputs may correspond to a large variation of one or more of the outputs. The linear approximations employed here [equations 23) and 24)] are poorly suited to follow such a behavior and Gibbs-type oscillations might appear in the solution, affecting the overall stability of the procedure and leading to numerical instability of the algorithm and divergence of the entire solution.

Transient simulation is a delicate task, and must be carried out with care [54]. But, what is of interest here is that the existing technology of process simulators is advanced enough to allow this task to be performed in an entirely automated fashion. A general flow chart for an unsteady solution method is shown in Fig. 2.9.

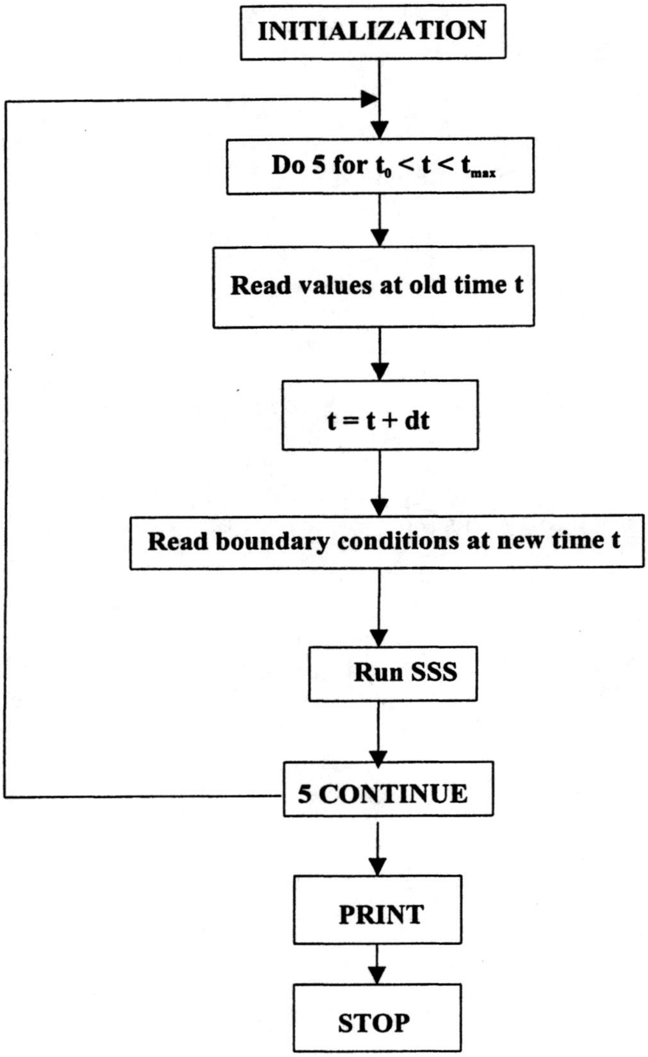

Figure 2-9. Flow chart of the unsteady procedure

2.4.2 - PROCESS SYNTHESISERS

2.4.2.1 - Artificial Intelligence and Expert Systems

In the preceding Sections it was -hopefully convincingly!- argued that *all* process design calculations can, in one way or another, be performed by properly implemented automatic routines. Process design is a highly labor intensive and highly interdisciplinary task and is therefore by necessity performed by teams of specialists: as a consequence, it is usually a very expensive task, and accordingly there is a strong incentive to reduce this labor intensity (measured in man-hours). The only task that has as of yet not been automated is the *conceptual* one: the choice of the type and of the characteristics of the process itself. In the following three Sections we discuss two different but clearly connected topics:

1. the possibility of constructing an automatic procedure capable of choosing the most convenient process configuration for a given set of design goals
2. the possibility of the actual implementation of this automatic procedure in a code.

This automation is facilitated by the application of the very powerful Artificial Intelligence (AI) techniques, whose specific task is to allow the codification of procedures which somehow mimic the thinking patterns of the human mind. We will actually make use only of a subset of these techniques, called Expert Systems ("ES" in the following), that can be used to reproduce the engineer's decisional path which proceeds from the design data and constraints to possible process configurations.

Expert Systems (see Section 3.1.3) are based on *relational languages* that use the symbolism of formal relational logic. They draw inferences from a number of *facts* stored in a particular database that is properly called a *knowledge-base.* These facts can be design data, design rules, physical or logical constraints, etc. Each ES manipulates this knowledge in its own way, according to a logical procedure contained in its *inference engine*.

AI techniques are in a sense the "main" but "hidden" subject of this monograph: they are all-pervasive throughout the entire book, but are never theoretically discussed or even explained in detail. Rather, they are presented here strictly from a user's point of view: readers interested in the theoretical and operational details of AI techniques are referred to some of the excellent monographic works on these topics [2,18,74] and for convenience to an introductory overview addressed to Process Engineers [48]. The only Process Synthetisers that have been developed in a verifiable working code are described in [5,31]. What is of interest here is that an ES can be constructed that performs in principle the following operations:

1) Acquires in machine-readable form a series of inputs that represent the knowledge of the state of the *universe* with which the ES will have to deal:

design data, type and state of the environment, component specifications, state-of-the-art of a certain technology, etc.;
2) Can be instructed by the user (with which it must interactively relate) about the logically and physically exact design goals, i.e. about some physical, logical or numerical properties the solution must possess;
3) Manipulates the data contained in its knowledge-base according to a predetermined set of rules contained in its *inference engine*, a sort of built-in logical flowchart that instructs the ES on how to proceed in different situations;
4) Can call other numerical procedures if necessary. Of interest here is the fact that an ES can call any process calculation code, like those discussed in Sections 2.4.1.1÷2.4.1.8 above.

2.4.2.2 - General knowledge representation for design applications

Let us assume that the decision has been made to construct an ES for a certain design activity. What are the necessary steps that lead the engineer from design inception (definition of need, general engineering concept, see [1,76]) to the generation and technical description of one or more final designs? This Section addresses this question. It is perhaps not unnecessary to offer a warning to the reader at this point: the more general the conceptual description of the design procedure becomes (i.e. the higher is the level of abstraction at which we attempt to describe the procedure), the higher becomes also the danger of overgeneralization. It is left to the reader to fill-in the gaps that may exist between the general method described here and any particular application.

The process of generating a design plan can be decomposed into three phases, roughly identifiable as *problem specification, functional analysis* and *design plan generation* [1]. We will present here a brief description of each of these three phases, and develop a detailed discussion in Chapter 4. In the first phase, we try to identify and define the (physical and logical) parameters that constitute the design goal, those that constitute the data, and those that constitute the constraints. In the second phase, we investigate the functional relations that exist among those parameters and how they must be accounted for during design. The third phase consists of a mapping of the logical design algorithm onto a relational procedure that can be implemented in an application using one of the available AI languages and tools.

2.4.2.2.1 - Problem specification

The problem specification phase is divided into two sub-phases:

a) **Problem identification and position.**

This is a task that can be performed on the basis of an itemized identification list:

- Can the problem be decomposed into a set of smaller and easier (in some sense) *independent* or *loosely dependent* sub-problems?
- From an opposite point of view, is it possible to consider the present design task as a sub-problem embedded in a higher-level problem for which a solution is already known?
- If the design problem consists of an assigned series of design steps, is it possible to ignore or skip those steps that are found to be either in conflict with others or superfluous for the general solution strategy? In other words, is it possible to establish *a priori* a hierarchy of importance of the different design steps, and rank them accordingly?
- Is it possible to foresee the complete problem domain (its *universe*)?
- Is a single solution sought after, or is it necessary to produce and compare several different solutions?
- Is the assigned (or gathered) knowledge-base self-consistent?
- Is it known from previous experience that the search for a solution requires a large body of knowledge, or is it sufficient to gather only problem-specific knowledge so as to somehow limit the solution space?
- Is the problem such that all of its knowledge-base is well-defined and unchanging, in which case the still-to-be-produced ES can run in batch mode, or is a portion of this knowledge-base uncertain or variable so that it is imperative to chose a highly interactive ES?

The answers to the above questions result in a set of guidelines that will indicate which methods are likely to be more useful in solving the problem and will direct actual implementation of the code.

b) **Identification of "always-relevant" parameters.**

The conceptual structure of the general process configuration must be analyzed and "decomposed" to derive a fundamental set of parameters and define an initial set of relations linking them to each other and to the design data. In many instances, this step is unnecessary, because the related knowledge already exists and has been codified in numerous examples of prior successful designs. If, however, this codification does not exist or is unknown to the user, it is necessary to consider that in any problem there are "always relevant" and "possibly relevant" parameters and relations. The first type includes all those that capture important and essential aspects of the functionality of the final design product and that are either always present in the basic set of similar design processes, or are deemed important by the Process Engineer at the time of launching the design activity. The second type includes all parameters and relations which are either important for particular cases only or whose importance depends greatly on the design specifications and in either case vary from problem to problem.

2.4.2.2.2 - Functional analysis: relations between the given specifications and the possibly-relevant parameters

The specifications (design data) may contain restrictions on a possibly-relevant parameter p_p, either directly (by setting a range of acceptable values for it) or indirectly (by setting a range of accepted values for another parameter r that is somehow correlated with p_p). In either case, p_p becomes *relevant;* and we will represent this by denoting it as p_r. It becomes necessary, therefore, to establish and investigate the chain of functional relations which link all the p_r to each other (by definition, all the "always relevant" parameters p_a are also p_r). These functional relations f are the usual design relations: equality, constraint (weak and strong), polynomial dependence, etc., and constitute a sort of unstructured skeleton of functional links between different parameters.

2.4.2.2.3 - Design plan generation

The Design Plan Generation Phase consists of the implementation of the skeleton plan produced in Phase 2 into a design plan. The relevant parameters are mapped onto (*set equal to)* the variables of a plan; the methods for assigning values are mapped onto (*developed into*) numerical or functional operators that act on the variables. The goals of the plan are then created, and the proper variable(s) are attached to them. Sets of variables corresponding to sets of tightly coupled parameters are attached to individual goals together with the applicable constraint(s). At the end of this process, all the elements of the design plan have been connected to each other in a network of goals that possesses (the network, and not necessarily the individual goals![11]) both the correct semantics of the original design problem, and the general rules of logical networks.

A design plan does not necessarily generate *a unique solution.* If the plan is correctly formulated, it will usually produce more than one feasible design. If needed, "optimization" procedures can be devised to choose, based on some criterion, the "best" one among different configurations proposed by the ES (but, see Section 2.3 above). Notice that, if we perform an optimization after having devised several different process layouts, *additional information is needed to execute this extra task.* This is always the case, in fact, when the only task we require of the ES is that of creating a list of feasible alternative process layouts: if

---

[11] There may be instances of conflicting goals (for example, minimum exchange area and maximum heat transfer in a heat exchanger) that must be handled by the design plan in such a way that they result in an "optimal" compromise (in this example, an optimal or sub-optimal heat exchanger configuration)

the logical patterns implemented in the code are correct, each one of the configurations proposed by the ES is feasible, but none is explicitly optimized at the time of its generation. Clearly, a far superior approach would be that of including some "logical" (qualitative) optimization in the process synthesis procedure: at the present state of the art, this is only possible for relatively simple and well-known processes.

2.4.2.3 - An example of automatic process design

Consider the most general problem that a designer can be faced with in the field of "energy systems": that of *choosing the most appropriate configuration to extract assigned amounts of exergy from a given resource base under some "environmental" as well as configurational constraints*. This apparently very complex problem turns out to have a relatively simple *qualitative* solution, but its complexity makes a *quantitative* solution difficult. This example is discussed here only to investigate whether it in principle admits a solution, i.e., whether it is possible to devise an as yet informal (= not coded!) procedure capable of completely solving the problem, robust in consistently reaching acceptable solutions, and effective in its search of the solution space. The problem is posed in the following manner:

A) Design data:

1 - The required exergy rate(s) to be "produced" must be specified both as to their *type* (electrical, mechanical, thermal) and their *amounts* (MW installed);
2 - The type of *environment*. This comprises the following information:
   - Type and amount of available resources (fuel, air, other materials that can take part in the process);
   - Amount of water available (qualitative information: "unlimited amounts", "limited amounts, but enough to act as heat sink for a Carnot process, "insufficient amounts", etc.);
   - Any environmental characteristic which can be foreseen to influence the choice of the process (e.g., "station is mounted on a barge", or similar);
   - All legal and/or technical constraints known to apply at the site under consideration.
3 - A set of *negative* instances, i.e. of processes that cannot be accepted as a solution (e.g., "nuclear plant not acceptable").

B) Knowledge about the available components which may be used in the process: the procedure ought to possess its own default library of units and sub-units which are normally employed in thermal plants. Such a list constitutes the knowledge base of the system about the tools it can avail itself to produce the required output starting from the prescribed input. A possible list [9] would look like the following[12]:

back-pressure steam turbine
brine heater
combustion chamber
compressor
concentrated pressure loss⊗
condenser
condensing steam turbine
contact (mixing) heat exchanger
cyclone
distillation tower
distributed pressure loss⊗
electric generator
electric motor
evaporator
feedwater heater
flow merger⊗
flow splitter⊗
fluidized bed combustor
fluidized bed heat exchanger
furnace
gas turbine
gas reformer
hydrostatic head⊗
mechanical gearbox
power merger⊗
power splitter⊗
pump
refrigerator
separator
stack
steam generator
surface heat exchanger
throttling valve

This list should be a dynamic feature of the code, in that it ought to be expandable by user-guided addition of new components. A component is identified by:

1. a set of input and output streams (specifying the kind of working fluid, the type of power it adsorbs or delivers, and identifying which output streams are the "main" streams and which are the "byproducts");
2. the set of "design parameters" that uniquely identify the component (inlet and outlet pressure, temperature, mass flow rate, power, etc.);
3. a set of mathematical rules that allow a complete specification of both the inlet and outlet streams given certain proper subsets of the design parameters.

A specific component is, in AI terms, an instance of a *class*. Such a class can be divided into *subclasses* according to some peculiar characteristic of each subclass. For example, if the *power range* is chosen, the class "gas turbines" can be considered as comprising, say, four subclasses: "less than 1 MW", "between 1 and 30 MW", "between 30 and 100 MW", and "above 100 MW". These subclasses exhibit all properties of the class to which they belong (they are said to have *inherited* these properties) but can have different values (or ranges) of the design parameters. Each subclass should be further divided (where feasible) into three types: "high-tech", "state-of-the-art" ("standard"), and "low-tech". Again, this allows for different values of specific design parameters (in the case of a gas turbine, for instance, inlet gas temperatures will differ).

The AI procedure ought to follow (or, as we said above, to *mimic*) the thinking patterns of the Process Engineer. The structure of these patterns is not difficult to realize: either a solution comes immediately to mind, or it must be

[12] Notice that some of the elements of the list (those marked by a "⊗") are not actual components. They must be introduced though as such, because they *enact a process* (decrease the pressure of a stream, or partition a flow rate or power, etc.).

sought after. Expressing this in a more formal way, we can say that there are two logical possibilities:

1. the problem admits a solution which is somehow "standard" in the mind of the engineer, i.e., it is so tightly and immediately associated with the formulation of the design task that it overshadows any other possible solution and even the idea of "thinking about" a different solution;
2. the problem does not admit of an "immediate" solution, and a logical chain of technical considerations must be deduced to reach a solution.

The first possibility is not of interest for the considerations that we will develop in this Section, but will be taken up again in Section 3.7, when we discuss the concept of "memory": it is clear in fact that this "immediateness" of the solution depends on both the experience of the engineer in the specific field and on his capacity of "recalling" from memory the necessary technical associations which make the solution "pop up". The point of interest of the second possibility is of course the "chain of technical considerations": while it is rather clear that the chain must end with the generation of a process layout, it is not immediately clear where it starts, i.e., what will be its first (logical) link. At the risk of overgeneralizing again, a little reflection shows that the chain works *backwards* (see Fig. 2.10), considering the design goal as an effect and trying to find its cause(s). Given the design goal, the engineer mentally scans all components he knows that would produce the given amount of thermal and/or mechanical and/or electrical power which has been recognized as the design goal. The first component satisfying the goal becomes the *last* component of a possible process P_1 that is still to be identified. If several components are found, there will be correspondingly several possible process configurations to develop concurrently. Then, the engineer turns his attention to the inputs required by the each one of the components just found, and mentally marks these inputs as "new design goals", and seeks other components whose outputs match each new goal. And so on, until:

a) all the required inputs are available in the "environment" (design specifications, point 2 above), and the configuration developed so far meets all specified constraints (design specifications, point 3 above). In this case, a process has been found, and the problem is solved. It is of course possible that many processes are considered feasible at this point, and all of them must be considered as proper solutions;
b) at a certain level in the chain of components, no component can be found that matches the required inputs under the specified constraints: the procedure is then aborted.

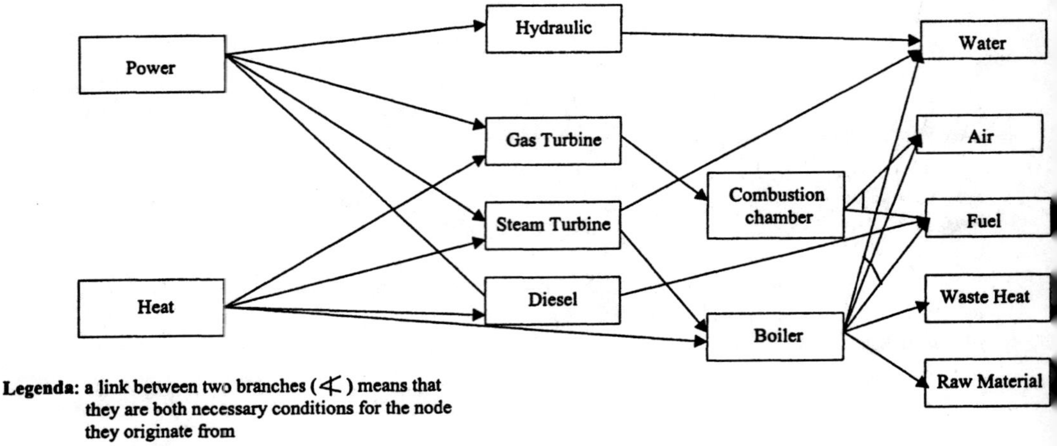

Figure 2-10. Inference network for power and heat production

To "match" a design goal means that the output of the component under consideration must be of the same type as the design goal (mechanical power, etc.) and that the range attached to that sub-class shares at least a sub-range with the range required by the design goal. There are cases in which only the first of these two conditions is met, i.e. in which the output range of the component under consideration does not match the required input. In these cases, the engineer would probably still consider the adoption of the component, but would take due notice of the mismatch, taking mental note that in the following of the design procedure some action must be taken to resolve this mismatch. Depending on the degree of the designer's experience, he would probably check at each step of this mental process whether there is the possibility that one or more "byproducts" or "secondary streams" which may have become available at previous steps (i.e., from components already successfully inserted in the provisional configuration) can be used to force the match (for instance, increasing the temperature of the water flowing out of the condenser to match the boiler inlet temperature). Additionally, an implicit "relative cost" concept is attached to each of the inserted components: the accuracy of this (not necessarily monetary!) "qualitative costing" will depend again strongly upon he degree of experience of the designer in the specific fields and on his general engineering knowledge. So, for instance, a nuclear steam power plant for electric power generation will be labeled with a higher installation cost than a conventional oil-fueled steam power plant.

This procedure can be translated into ("mimicked" by) a computational procedure, written in the language of *relational logic* (some symbolic or object oriented language, see Chapter 3), as follows. Notice that the conceptual similarity between the human and the machine deduction processes leads to striking similarities in the description of the activity as well:

Given the design goal, the inferential engine *scans* the *component library* to seek the components that would *meet the goal* (which is a given amount of thermal and/or mechanical and/or electrical power). The first component satisfying the goal is placed in the first slot of the *working memory* as the *last* component of a possible process P_1 that is still to be identified. If n components are found, there will be n possible "*process trees*" on which the procedure has to work. Then, for each P_i ($i = 1,...n$), the inputs required by the component just found are taken to be the *new design goal*, and the library is scanned again seeking for components whose outputs match the goal. The procedure is repeated until:

a) all the required inputs are available from the class "environment" and the configuration developed so far meets all specified constraints. In this case, a process has been found, and is displayed to the user in a proper form (if more processes have been found, *all of them* are displayed);
b) at a certain level in the process tree no match can be found (under the specified constraints) for the required inputs. The procedure is then aborted.

If the procedure identifies a mismatch, it tries to work around it by keeping track of all mismatches and checking at each step about the possibility that one or more "byproducts", or "secondary streams" can be used to force the match.

Cost considerations can of course be explicitly included in the code, which has a much higher capability of executing concurring calculations than the human mind. In addition, a "relative cost" can be attached to each of the constructed processes, which is a function of the number of components and of the number and amount of external resources used. So, for instance, a steam power plant with an externally fueled feedwater heater train will have a higher "cost" than the same process with regenerative feedwater heating[13].

2.5 - PROCESS OPTIMIZATION: THE CLASSICAL VIEWPOINT

Referring the reader to the definitions and to the considerations introduced in Section 2.3 above, we shall try here to describe in some detail what is currently meant by "Process Optimization". To begin with, let us define the problem in practical terms, by examining the questions that must be answered by an optimization procedure. They are exemplified by the following list [1,12,23,49]:

---

[13] To avoid possible confusion, it must be stressed here that this "cost" may *include* monetary cost as well but is in general to be regarded as a sort of *weighting* or *penalty function* whose value is used by the search algorithm of COLOMBO to scan the decision tree more efficiently.

a) is there only one physical process that produces the required output, or are there more than one?
b) for each transformation which composes a process, is there only one piece of equipment capable of performing it, or are there different types or classes of components that perform similar processes?
c) what is the most convenient process structure, i.e., how can components be arranged (connected) under the given constraints to obtain the best results in terms of the overall objective function?
d) what is the most convenient size of each component?
e) what is the most convenient technological level of each component?
f) what is the most convenient physical location of each component in the plant layout?
g) what is the most convenient value of each one of the operating parameters?

It is important to realize that these questions do not generally admit of univocal answers: this leads to the very important consideration that *however well formulated and "exactly" solved, an optimization problem possesses an intrinsic qualitative essence that cannot be captured by a purely deterministic approach.* We shall return to this point in Section 2.6 here below. To remain within the realm of the classical viewpoint, let us recognize first that for what Thermal Systems are concerned, it is necessary to distinguish between two levels of optimization, the *structural* and the *operational one*: structural optimization attempts to find the "best" possible process configuration that achieves the specified goals under the given constraints, while operational optimization tries to find the most convenient combination of design parameters so that a given process configuration attains "best" results. Though not acknowledged at all in the pertinent literature, it is clear that the two procedures are *logically at different levels* (the structural being at a meta-level with respect to the operational): as a consequence they can be run in sequence, possibly under an iterative protocol. A first configuration could be computed and the optimal values for its operating parameters calculated; then a structural modification could be introduced, and the operational optimization repeated, and so on, until a global optimum is reached. In principle, such a protocol is straightforward, but it suffers from two main limitations: it is computationally very taxing (because it consists of a series of nested optimization loops, one for each configuration); and -more important- it requires by necessity (at least in the formulation presented here) the scanning of the entire process tree. In fact, once the first configuration has been optimized, no clue can "automatically" be found to guide the designer in his search of some structural modifications to produce a second configuration. This decision is in fact invariably taken on the basis of the designer's experience and knowledge of the process phenomenological description (i.e., of the mental model he has of the process!). While the fast advancing hardware technology can solve the first limitation in the short run, there is a conceptual problem at the basis of the second one: structural and operational optimization may share the same objective function, but their domain of application (the decision variables), the system boundaries and the solution space are different. For an operational optimization, the decision variables are the technical and economical system parameters, the system boundary is the physical boundary of the given configuration (which is fixed), and the solution space contains all of the possible sets of independent technical and economical design parameters. For the structural optimization, decision variables are the different possible configurations, the system boundary may vary from a configuration to another, and the solution space consists of the set of all feasible configurations.

As a result of this state of affairs, virtually all optimization procedures available in the literature are of the operational type, the structural optimization is left to the ingenuity of the process engineer, and no codification is offered for it. The only and remarkable exception is the so-called *Simulated Annealing* optimization [53], in which structural optimization is embodied, though without a high level systematic decision-guiding protocol.
These procedures invariably consist of the following steps:

1. *problem formulation*: definition of the system to be optimized, of its boundaries, of the independent and dependent variables, of the constraints;
2. *formulation of the objective function*;
3. *problem formalization*: writing of a (generally ad hoc) numerical code to implement the optimization procedure; the code may make use of one of the well-established standard numerical optimization libraries (like IMSL, MATHEMATICA, MATLAB, etc.)
4. *problem solution*: application of the code to the specific problem under study;
5. *critical review*: analysis of the results and possibly reformulation of the objective function;
6. *possible iteration* from 2 to 5 (in Simulated Annealing, from 1 to 5).

A protocol based on points 1-6 above is highly labor-intensive: in fact, steps 1,2,3 and 5 require a substantial investment in terms of man-hours at specialistic level, and possibly only step 3 could be eased by the recourse to existing general optimization packages. At present, complete optimization of all design parameters is routinely run only on isolated components; approximate optimization (i.e., one in which the set of independent variables has been strongly reduced and/or redundant constraints have been added to limit the size of the solution space) is carried out for quite complex systems like combined-cycle plants [12], gasifier platforms [1], entire chemical plants [77], etc.. A technique often used to reduce the required amount of computational resources is that of performing a certain number of *sub-optimizations*: in this approach, certain portions of the plant or process are considered in isolation, and proper additional boundary conditions are imposed to take into account the interchange with the remaining portions. This procedure rests on the assumption that the performance of the sub-system is substantially the same both when considered in isolation and when inserted in the "context" of the overall process: provided the additional boundary conditions are properly chosen, this condition can always be satisfied. Unfortunately, the "optimal" operating point of a sub-system may be substantially different if its outputs are considered *per se*, or if the same outputs are to be considered as "internal" fluxes to a larger system: therefore, this technique must be used with care. We shall not delve further in this topic, and refer the interested reader to specialistic treatises on general optimization procedures [53, 57], or to specific textbooks that deal with Thermal Systems Optimization [1,14,76].

2.6 - Some Additional Remarks on the Optimization of Thermal Systems.

The "classical" approach to optimization, in spite of its invaluable practical usefulness, is to be regarded, as already indicated in Section 2.3, limited and in a theoretical sense obsolete. This may seem, and in fact it is, a very strong statement, and is used here to mean that, while as a tool classical optimization methods can still be employed, time is ripe for exploring some other avenues to "optimal design". Classical optimization methods suffer of a fundamental internal contradiction: while they are constantly and explicitly praised for the "exactness" of their procedures, their purely deterministic approach makes them incapable of capturing either the meaning, or the goals or even the concept of "optimization". One must recall a very important point: no matter how apparently exact the formulation is, and how accurate the computational approach is, *an optimization problem does generally admit of more than one solution, and often these solutions show qualitative rather than quantitative differences.* This fact has very important consequences on the rational design of an actual optimization program, but has been routinely overlooked in the literature (with notable exceptions, see [44,64]). But there is an even more important fact that we want to stress here: *when using their experience and knowledge to solve an optimization problem, Process Engineers heavily rely, consciously or unconsciously, on qualitative much more than on quantitative thinking.* This observation, coupled with the large extension of the solution space, leads to the conclusion that optimization *per se* ought to be performed as a fuzzy process, one in which both the goals (the objective function, the "thinking") and the means (the optimization procedure, the software) are not susceptible of "exact" quantification. Of course, this does not mean that the problem formulation cannot be cast in an exact mathematical form: it indicates rather that, if the minimum of the objective function is for example attained for "p_{max} = 1.8 MPa, T_{max} = 750 °K", it is likely that the *real* optimal solution will lie in a neighborhood of these values ("p_{max} = 1.8 ± 0.05 MPa, T_{max} = 750 ± 25 °K") rather than being constrained to exactly abide by the calculated numerical values of the independent variables. Furthermore, the limits of fuzziness, i.e., the "± 0.05 MPa" as well as the "± 25 °K", may be determined by a qualitative (approximate) reasoning rather than by exact mass- and energy balances. Let us reconsider the general questions that constitute "the optimization problem" at large (Section 2.5), and see where a qualitative approach is of importance:

a) *is there only one physical process that produces the required output, or are there more than one?*

This requires some analogic thinking: electrical power can be generated by a Diesel genset, by a turbogas- or steam-turbine driven generator, by an array of photovoltaic cells, by a hydroelectric plant, by an array of fuel cells, etc.: while some of the choices can be pruned (discarded) by considering quantitative design specifications or boundary conditions (e.g.: "no steam plant below 1 MW", or "no Diesel above 100 MW" , or "no hydroelectric plant without a suitable water reservoir", or "no fuel cells if the cost of the generated energy must be lower than 0.1 $/kWh", etc.), others are in fact equivalent in principle, but can be discarded or accepted by using some qualitative reasoning, like "Diesel engines require relatively little maintenance" or "turbogas gensets are plants of high technological complexity", etc.. The point is, the "experience" of a designer is more often qualitative and analogic than quantitative and deterministic.

b) *For each transformation which composes a process, is there only one piece of equipment capable of performing it, or are there different types or classes of components which perform similar processes?*

Same considerations as above, with one addition: while the "main" transformations are in general well defined, and there are in principle certain components that are by their own nature suited to perform them (compression → compressor, combustion → burner, expansion → turbine, etc.), there are "secondary" or "accessory" transformations that can be realized under slightly different conditions by different types of components. One example is the reduction of the pressure of a gas: this can be achieved by throttling (valve) or by expanding (turbine), and the choice depends on the pressure ratio, on the mass flow rate, on the importance of the transformation in the general economy of the process, etc.. Another example is the heating of a stream, which can be obtained by introducing an *ad hoc* thermal source (an auxiliary burner) or by direct or indirect coupling with a hotter stream (mixing or surface heat exchanger): the choice often depends on the known degree of success of the two strategies, and this knowledge constitutes the "experience" of the designer.

c) *What is the most convenient process structure, i.e., how can components be arranged (connected) under the given constraints to obtain the best results in terms of the overall objective function?*

Again, there is a high degree of qualitative thinking implied in this decision: memory of past designs, analogy with successful configurations, or simply good thermodynamic reasoning can guide the selection much more efficiently than purely quantitative considerations. Notice that Simulated Annealing procedures can solve "exactly" this structural optimization, but they do so following a rigid protocol, and are therefore not "intelligent"; as a side remark, they are also very computationally intensive, because they must evaluate in sequence *all* of the possible configurations, starting from the most complex one.

d) What is the most convenient size of each component?

This question seems the one more apt to a simple quantitative answer, and it generally is: but there are instances in which qualitative considerations are very effective in the generation of proper designs. Consider for instance a problem frequently encountered in HEN design (see Example 6.9), namely that of thermally coupling two streams with very different heat capacities. It is well known in practice that it is convenient to split one of the streams, so that the coupling takes place between two streams of approximately equal heat capacities: but it is difficult to generate a quantitative procedure to enforce this rule; so, stream splitting is more conveniently introduced and controlled by a approximate reasoning approach.

e) What is the most convenient technological level of each component?

Again, the answer is dictated by approximate analogic thinking, and not by deterministic reasoning.

f) What is the most convenient physical location of each component in the general plant layout?

Though nobody would ever locate the condenser extraction pump far away from the condenser, or the steam attemperator underneath the LP turbine, and though both economic and thermodynamic quantitative considerations can be developed to justify each decision, in essence the motivation is *analogy with existing plant structures known to perform satisfactorily*. This is clearly a case of *memory retrieval by matching* (see Section 3.7): the location of the extraction pump is known for all condensers built in the past, and it happens to be underneath the condenser itself: the fact "the plant contains a condenser" becomes the parameter for which the match is sought, and "the position of the pump" becomes the parameter to be recalled from memory.

g) What is the most convenient value of each one of the operating parameters?

This is in fact a mostly quantitative decision, and the only qualitative attribute that can influence the choice is the technological level of the component.

We hope it is clear from the above considerations that a rational approach to optimization problems requires that the optimization be handled qualitatively as well as quantitatively, and that the qualitative reasoning be introduced at a "high" level into the design procedure (that is, in the early stages of the design activity). As it was mentioned above (Section 2.3) design and optimization are in reality carried out together in the mind of the process- and design engineer, and ought to maintain this close link in any AI-based design paradigm.

In the next two Sections, we examine two optimization techniques that by their own nature are very easily adaptable to this design-by-optimization philosophy.

2.6.1 - THERMOECONOMICS

Traditional engineering wisdom instructs the Process Engineer to search for "the most economically convenient solution". This is the basis for all optimization procedures, of course: the overall cost per unit of "product" ought to be the lowest permitted by the design specifications and the technological and economic constraints. It is these "economic optima" that have been sought after for years, and components, processes, plants and entire production platforms have been constructed according to this strictly monetary performance requirement. There may be instances in which the criteria for success are not purely monetary (see Sections 2.1.1.1 and 2.1.1.2): in such cases, a "monetary value" has to be assigned to the non-economic parameters to ensure the proper homogeneity of the quantification. This type of economic optimization usually leads to a formulation of the objective function in terms of "minimum unit cost of the products", or -which is equivalent- of "maximum rate of return (ROI, Return On the Investment)", or "minimum pay-back time" [1,76]: from its systematic application a complex, universal and well-tested series of procedures have been perfected that provide "exact" answers, in the sense that they can calculate the optimal value of a set of relevant parameters with any reasonable degree of accuracy, and that they lead to reproducible results. This "monetary" approach has of course been successful to a high degree, because our "Western" industrial society has thrived, the efficiency of our conversion devices has increased, and the overall "quality" of the manufacts has constantly improved in the last centuries: but the obvious success on the large scale has contributed to hide three important drawbacks. First, there are so many uncertainties in the assumptions necessarily introduced by the process of formulating a well-posed problem, that the "exactness" of the solution, that is, its "optimality", is questionable in line of principle. Second, the approximation introduced by assigning a monetary value to non-economic parameters (like for instance social or strategic considerations) introduces an additional degree of uncertainty in the solution. Third, all economic optimization methods fail when they are used to assign monetary values to distinct products of the same process: this is often referred to as the "by-product cost problem", and applies to any process having at least two outputs of different energy quality[14] (e.g., steam and power, or chilled water and waste heat, etc.). In these cases, it is necessary to apportion the capital and operational costs between each one of the outputs: it can be shown that, if the apportioning is made on the basis of the energy content of each stream or product, and even if the additional assumptions necessary to perform the apportioning are made on

[14] The *quality* of an energy flux is a *non-quantitative attribute of the object "energy"*: it can be measured quantitatively by the fraction of this flux that can be converted into mechanical work by means of reversible transformations that bring the flux in equilibrium with its surroundings. A (mechanical or electrical) power flux has a quality of 1; steam or gas flows at a temperature T_h have a quality equal to $1 \pm \frac{T_o}{T_h}$, if T_o is the temperature of the environment.

the basis of an accurate evaluation of the real costs associated with each component of the plant, the cost of the "by-product" (i.e., of the stream with the lowest energy quality) is always overestimated [16]. The first of the above problems has been already discussed in Section 2.6, and calls for a different approach to optimization at large. The second and the third drawbacks can be solved by using not the energetic but the exergetic content of a stream or a product as a cost basis. In the last decade, exergy-based optimization methods which formulate the objective function in monetary terms have been perfected [1,12,36]: as a matter of fact, they have been so successful that a new name has been coined for them, *Thermoeconomics*, or "*exergy-aided cost minimization*" [1]. A Thermoeconomic optimization consists of the following steps:

1. calculate the exergy destruction rates and the exergy losses for all components of the system;
2. associate a monetary cost to each exergy stream (which can be a flow of energy or of matter);
3. compute the unit costs associated with each component and due to capital investment, operating expenses, maintenance, etc.;
4. assign a monetary cost to events possibly occurred outside of the system boundaries, but that may be reflected in the operational costs. This is called *internalizing externalities* (for instance, consider as a cost internal to the system the cleanup of possible pollution caused far away from the plant);
5. derive an expression for the unit cost of each product. These costs can be combined to obtain the overall production cost, which is then the objective function;
6. for a fixed system configuration, modify some operating- or design parameters and compute the changes in the value of the objective function. Proper use of standard mini-max techniques ensures the attaining of an extremal point.

Thermoeconomics still suffers from the first of the drawbacks mentioned above: it can operate only on a fixed system configuration. Though in principle feasible, no attempt is known of a "Thermoeconomic Simulated Annealing", which could perform some structural optimization.

2.6.2 - EXERGO-ECONOMICS

Proceeding on the same line of thought as Thermoeconomics, and going one step further, it has been argued that it is possible to avoid the attribution of a monetary cost to units of energy (or exergy) and *perform the optimization on the basis of the exergetic costs only*. This *exergetic cost* is defined as *the net exergetic content of a stream or physical product*. While the exergy of a stream is easily computed, there may be some questions about the exergetic value of the unit mass of raw materials entering a plant, or about the exergetic "value" of the final products. The solution is found [21,36,77] by the following considerations:

1. the exergetic cost of a stream is the product of its specific exergy content multiplied by its mass flow rate;
2. the specific exergy content of a unit mass of matter or of product is the sum of all the exergies of the streams that have been used in the chain of technological processes leading to its extraction or production;
3. the exergetic cost of an externality is the amount of energy and materials (measured each by its exergy) necessary to "resolve" the externality (for instance clean-up costs, or dismantling costs for obsolete plants, etc.).

From the computational viewpoint, the advantages of using an Exergo-Economic instead of Thermoeconomic optimization are marginal[15]: the most relevant one being the fact that there is a conceptual homogeneity in the terms that make up the cost balances, and that exergy fluxes can give an immediate graphical representation of the conversions performed by the system. The importance of Exergo-Economics is conceptual, in that it constitutes an optimization of *values* and not of costs or prices, and as such is more suitable for macro-economic considerations, in which monetary issues are of secondary importance with respect to sustainability or social issues. Some Authors use the two terms "Exergo-Economics" and "Thermoeconomics" as synonyms [1,43], but they are not, and the distinction having its own merit ought to be maintained. Thermoeconomics introduces in our optimisation procedures the concept of *cost*, intended as *price* of a commodity or of a service, and shows that by inclusion of this cost into the set of relevant performance parameters, a "most convenient design configuration", or a "most convenient operational point" can be calculated which are different from those derived from purely thermodynamic considerations. Exergo-economics also expands on thermodynamics, but considers as the additional relevant parameter the *exergetic cost* of a commodity or of a service, which is a measure of *value* and not of price [21, 67]. The difference is not trivial: while a Thermoeconomic analysis performed on a certain process configuration is subjected to the price oscillations dictated by the economic markets (for the commodities and services needed in that process), an Exergo-economic analysis of the same process configuration maintains its validity in spite of the volatility of the price structure. Of course, such an analysis would not be of eternal validity, because even the values of commodities and services can be modified by the discovery of new raw materials, or by more efficient extraction processes, or by refined technological pre- or post processing technologies: but it is clear that Exergo-economic analyses are subjected to at least one less uncertainty parameter, namely price. In conclusion we can say that Exergo-economics *embeds* Thermoeconomics!

[15] It must be mentioned though that to perform an Exergo-Economic Analysis all the exergetic costs must be known: this requires the acquisition of a large Knowledge Base both on the numerical values of the exergies of raw materials in Nature and on the exact chain of technological processes that leads from each "raw" material to the "refined" or "finished" fuel or auxiliary product [77,78]

From the point of view of qualitative optimization, the importance of Exergo-Economics and Thermoeconomics is threefold:

1. they introduce a uniform criterion for evaluating all fluxes of matter and of energy;
2. they allow the assessment of the conversion efficiency of whatever process by a consistent figure of merit (the exergetic efficiency);
3. exergy-based reasoning can be systematized into a set of relatively simple rules that can be then taken to form the basis for an Inferential Engine performing qualitative optimization [16,64,65] (see also Examples 6.7, 6.8 and 6.9)

CHAPTER 3

ARTIFICIAL INTELLIGENCE

3.0 - ARTIFICIAL INTELLIGENCE IS NO INTELLIGENCE!

What is intelligence? Or, putting the question in a form more suitable to the paradigms of propositional calculus: *what are the attributes a (not necessarily human) being must possess to qualify as "intelligent"?*. There is considerable debate among specialists about an exact definition, but one which would be accepted by the majority of modern cognitive psychologists is the following [6,69]:

An individual can be said to be intelligent if he/she:

1. *reacts in a flexible way to different situations;*
2. *is able to take advantage of fortuitous circumstances;*
3. *can correctly assess the objective relative importance of the different elements of a new situation he is faced with;*
4. *can solve with original (i.e., hitherto unknown to him) means a problem he has no previous experience of;*
5. *can (deceptions aside!) correctly understand thinking patterns different from his own;*
6. *is open-minded, i.e., he is ready to accept the possibility of being wrong both in his causal and in his "intuitive" thinking;*
7. *can produce new and innovative ideas.*

(Both Authors regretfully admit that they would not count as "Intelligent Beings" according to the above definition!).

One does not need to be a computer expert, or a logician, or an AI practitioner, to see that *at least in the way we conceive computers and computer languages today, there cannot be an "intelligent android"*: however complex his hard- and software, it (the android) will in all likeliness lack (at least) qualities 4) and 7) in the above list[1].

A remark first expressed in [21] can help us in putting things in a perspective closer to the context of this book: in common language, we attribute the quality of "intelligence" both to individuals *who must think because they do not know much*, and to those *who know much, and therefore can afford to think less*. In the terms of AI, this can be rephrased as follows: we tend to attach the attribute "intelligent" both to individuals who possess a small amount of Domain Knowledge and must therefore exercise a large quantity of Inference to draw their conclusions, and to individuals who conversely possess a large amount of Domain Knowledge and can therefore reach the same conclusions by performing a rather limited Inferential activity. Notice that both in the first and in the second definition we have implicitly assumed that an *intelligent reaction is neither spontaneous nor necessarily dictated by the available data.* This point is important for two reasons: first, it makes a clear distinction between "potential" intelligence, intended as a "memorization of collected knowledge", and "dynamic" intelligence, which is at once the act and the result of "reflecting" upon the collected knowledge. Second, it reminds us that all concepts about "thinking" reported in the following of this Chapter refer to *human thinking modes somehow hard- or soft-coded into a machine*, and that a one-year old human (or, for that matter, a one-month old cat!) can "think" (= "infer") and "learn" (= "increase the amount of Knowledge he commands on") much faster and more effectively than today's most advanced AI programs. On the other hand, it must also be acknowledged that AI methods give origin to codes that perform with almost incredible reliability and speed cognitive (qualitative) tasks which mimic very closely some traits of human thinking patterns: if we agree with the idea that potential and dynamic intelligence are operationally equivalent (in the sense that they can reach the same results), then it is clear that there is no limit to the "IQ" of an AI code, because all it takes to increase it is to store more (properly constructed and connected) knowledge in its Data Base. In this sense, computers *are* intelligent even for today's standards.

---

[11] We must mention here that there are AI researchers that would strongly disagree with our statement: in fact, there are examples of AI programs that have "discovered" new (albeit admittedly secondary) "laws of Nature" [20], and it would be difficult even for an expert to see where Deep Blue (the Chess Computer Master by IBM) lacks "intuition" in its chess moves [20]. But, the point we want to make here is somewhat less general, and limited to the realm of engineering design: within this realm, it is (yet?) unconceivable that an AI program "discovers" Fourier's law, or "invents" the Carnot cycle.

3.1 - DEFINITION OF CONCEPTS AND TERMS

3.1.1- ARTIFICIAL INTELLIGENCE ("AI")

There is no universally accepted definition of AI. One which is sufficient for the purpose of this Chapter is the following:

AI is that part of Computer Science that investigates symbolic, non-algorithmic reasoning processes, and the representation of symbolic knowledge for use in machine inference.

This is a rather modern [3] and quite complete definition, which contains more than one point of interest for the following of this Chapter. Firstly, it stresses the symbolic character of the object of study of AI. Though its meaning is obvious, its implications are not so: *AI's field of interest does not contain numbers and operations on them, but logical objects and their mutual relationships*. Secondly, it introduces the concept of *non-algorithmic reasoning*, which is usually very difficult to grasp for an engineer: a solution to a problem is not obtained by a pre-definite series of operations, but rather by a series of logical inferences, consisting in an *a priori* unknown number of steps, and whose outcome is in no way predictable in its entirety. Thirdly, this definition correctly states that scope of AI is to generate a process of machine-implemented inference, i.e., to construct a computer code which can infer a *cause* from an *effect*, possibly with the help of some additional (logical) constraints.

Historically, the first reference to the concept of "Artificial Intelligence" is generally credited to McCarthy [71], but his original definition was quite broad and gave origin to a number of misinterpretations and misconcepts which still burden this branch of Computer Science:

AI is that compendium of computer techniques dedicated to the implementation of a simulating procedure for any mental process.

Today, this definition seems too ambitious, because as our studies in the cognitive sciences (Psychology and Bio-Neurology) gain in depth and breadth, it appears that many of the schemes adopted in the past to explain mental artefacts like memory, selective memory, learning and understanding are only crude approximations, or better yet primitive representations, of the complex chains of bio-chemical phenomena taking place in the brain. Therefore, the idea of simulating "any mental process" has been replaced by the following assumptions:

1. "Intelligence" can be explained and represented as a symbol-manipulating activity;
2. this activity can be embodied in a physical symbol system; in particular, it can be both *described* and *implemented* in such a system;

3. the symbol manipulation necessary to the description and to the implementation of any "intelligent activity" can be carried out on digital computers;
4. various aspects of human "intelligence" (in particular, what is usually called "logical reasoning" in common language) can be modelled by such physical symbol systems;
5. there may be a "general theory of Intelligence" that may originate a "general symbol system" which in turn might result in a "universal computer code" capable of describing all "intelligent phenomena": however, even if its discovery should be a long range goal of AI research, its quest ought not to go to detriment of "partial" or logically "local" symbols systems (which result in "specific" applications)[2].

Recently, another purpose for AI has been put forth: as noted in [41], in spite of some humbling failures that helped bring things back into a more down-to-earth perspective, AI methods and devices can "serve as windows on the mind", i.e., help develop a scientifically pragmatic understanding of the very concept of "Intelligence".

3.1.2- KNOWLEDGE

To be able to "handle" knowledge (i.e., to put it to use), one has to make a distinction between two types of knowledge:

1. *Declarative knowledge*, i.e., knowledge of *facts* and *relationships*; also called "knowing what". Examples:
a) a turbine is a rotating, bladed wheel that absorbs energy by interacting with a fluid;
b) a desalination process is a set of components that operate on an input stream of seawater and, by making use of some external energy source, separate a large portion of the dissolved salts in the water;
c) burning coals with high ash- and sulphur content induces severe fouling in the boiler tubes and produces acid stack gases.
2. *Procedural knowledge*, i.e., knowledge of the cognitive steps to take to *draw certain conclusions from a data set*. Also called "knowing how". Examples:
a) to decide whether a machine is a turbine, check whether:

---

[2] The reader must be warned that this seemingly "technical" dispute (about the possibility of simulating "thought" tout-court) has become a phylosophical issue between those who negate the possibiity of "strong AI" and underline the unicity of the human mental activity [52], and those who advocate a possible future possibility for the implementation of a "universal virtual simulator", undistinguishable from a human brain [6]. The topic obviously exceeds the limits of this book: for a better comprehension of the essence of AI methods and for a (not always impartial!) discussion of their possible implications, it is useful though to consult [20]

- it is rotating;
- it has blades;
- it absorbs energy by a fluid flowing through it;

b) to generate potable water, one possible solution is that of:
- pumping seawater into a desalination plant
- provide the desalination plant with some sort of external energy input (heat or work)

c) if fouling in the tubes of a certain coal-fired boiler is higher than usual, check the history of the quality of the coals that have been burned in it;

d) if stack gases are high in SO_2, one can either de-sulfurize the gases in a De-SOx unit, or de-sulfurize the coal before it is fed to the boiler.

In spite of the above, "knowledge" is a difficult concept to define. Quoting from [63]: "*It is difficult to define knowledge, difficult to decide whether we have any knowledge, and difficult, even if it is conceded that we sometimes have knowledge, to discover whether we can ever know that we have knowledge in this or that particular case*". But we are engineers, not philosophers, and are interested in both *knowing-how* and *knowing-what*, so that we can differentiate various types of knowledge that engineers avail themselves of [74]:

- knowledge about *Objects* (plant, process, component, chemical compound);
- knowledge about *Properties* (weight, material strength, Lower Heating Value, size);
- knowledge about *Activities* (drawing, selecting, calculating, building, assembling);
- knowledge about *Time-scheduling* (design precedes construction which precedes testing which precedes operation);
- knowledge about *Aggregations* (boiler composed of burner + tubes + drum + economiser);
- knowledge about *Heuristics* (T_{max} of current technology gas turbines is about 1300°K if uncooled, 1500°K if cooled);
- knowledge about *Causality* (High SO_x emissions caused by high-S fuel);
- knowledge about *Proper Assumptions* (differential thermal expansions can be taken care of by proper design);
- knowledge about *Procedures* (design- or testing protocols);
- knowledge about *Quantitative facts* (average LHV of lignite is about 27000 kJ/kg).

In conclusion it can be seen that under the name of "Knowledge" we refer to the complex of information (relational, numerical, topological, etc.) which constitute the data-base in which an AI procedure searches for solutions. What can be included in this "Knowledge" will be made clearer in the following of this Chapter: the list above implies that it can include both *data* and *procedures* (even *calculations*!), and both *deterministic* and *probabilistic* links.

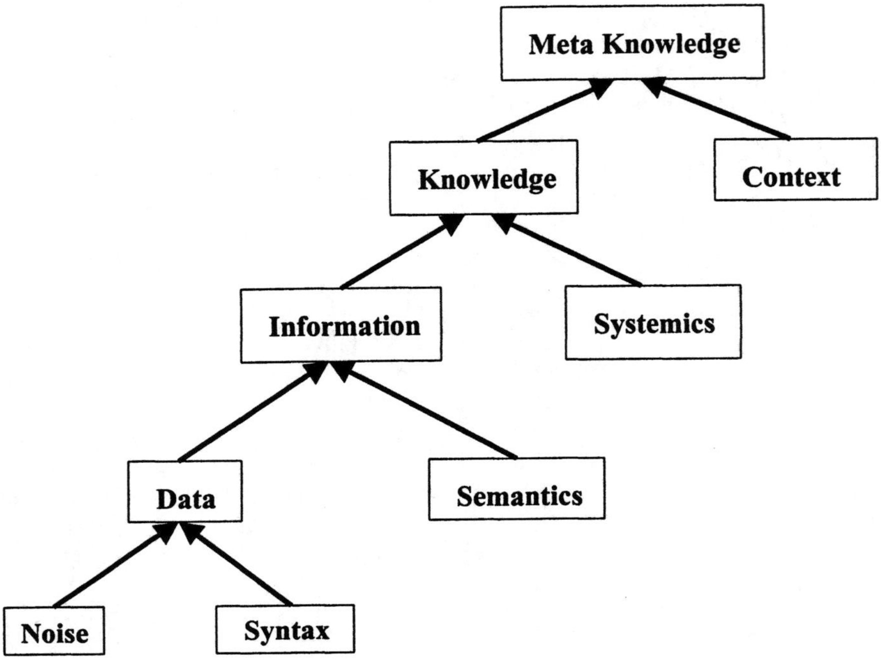

Figure 3-1a. The path towards "understanding"

As a final remark, it should be always kept in mind that knowledge is by its own nature hierarchical: consider the graph of Fig. 3.1, which depicts our path towards "understanding" a problem (in this case, the fact that burning coal produces heat). The physical world (our primitive "universe of knowledge") presents to us a large set of *randomly available, noisy, casual and unorganised information* (the fact that coal may self-ignite, that it can be gasified, that it develops a certain amount of heat per unit mass, that we can somehow collect this heat flux and put it to use, etc.), which we can filter through our common technical sense and through the rules of a *domain-specific syntax* to obtain a set of "catalogued" data: these data do not possess by necessity any *a priori* logical structure, but rather have been collected according to some reproducible criterion (for instance, classifying the coal by its type and chemical composition, stating the necessary conditions for its ignition, those for a steady combustion, for collecting the heat flux, etc.) .

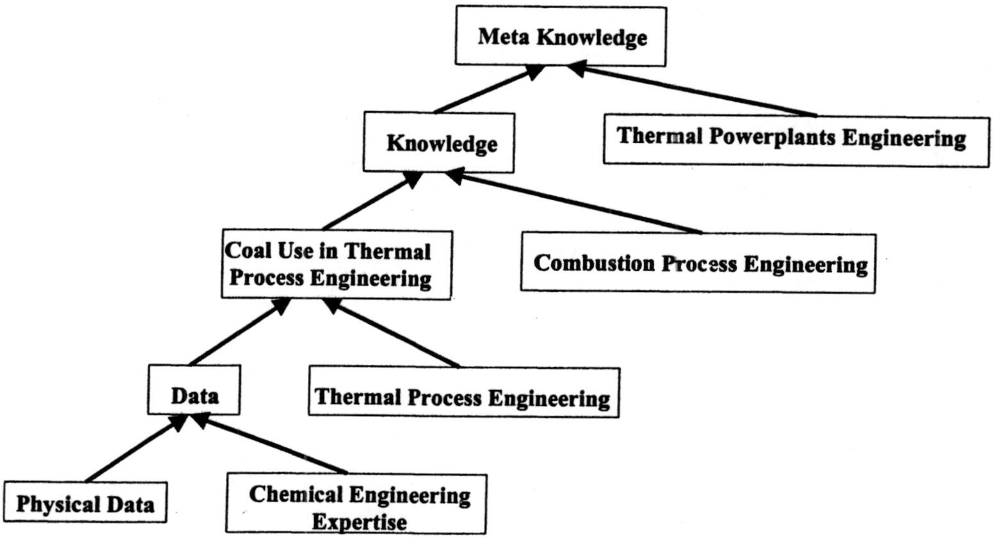

Figure 3-1b. An example of application of the logical procedure of Fig. 3-1a.

Operating on these data with a proper *semantic* (which does not need to be domain-specific [i.e., related to the coal], but is usually at least *field-specific* [in this case, related to Thermal Process Engineering]), we collect a number of information bits, which can be used to build a certain amount of domain-related knowledge if we operate on them with the help of some domain-specific logic, which can be called "*systemic*" or "knowledge about the system" (to continue with the description of our example, Combustion Process Engineering). Finally, this first-degree knowledge can be re-evaluated, reorganised and systematised by examining it in a larger *context* (that of Thermal Powerplants, so that for instance we "understand" that coal can be substituted by other fuels of different physical and chemical structure): we reach a *meta-knowledge* about our domain, on which we could again operate an abstraction, i.e., put it in an even larger context (co-generation of heat and power), and so on. Any automatic knowledge-handling paradigm devised to model ("reproduce") the entire knowledge build-up procedure must therefore be hierarchical in its structure.

3.1.3- EXPERT SYSTEMS ("ES")

An ES is an application (i.e., a series of procedures aimed at the solution of a particular class of problems) built around a direct representation of the "general expert knowledge" which can be collected (and not necessarily systematised in any way) about those problems. In general, an ES associates a "rule" (see below, 3.1.6) to every "suggestion" or "advice" a specialised expert would give on how to solve the problem under consideration. The name *Expert System* is inaccurate and misleading, because it carries a sort of mystical flavour. The systems we are going to deal could be better described as *Knowledge Based Systems* ("KBS"). The class of KBS (or ES) we describe in this book do not use the same thought process as humans: *they do not think*, nor do they really mimic human expertise. They simply store, in a special type of database, a rather limited amount of knowledge about some specific features of the real world which is put into them by human "experts", in such a way that it can be formally manipulated, used to infer results and, in general, represented and applied in an orderly fashion when and as needed. An ES is as trustworthy as the information supplied to it by the expert(s). One of its inherent advantages, though, is that it can be interrogated about its reasoning, so that false and wrong knowledge can be easily detected, and cannot be concealed in obscure computer code. From the “structural” point of view, a KBS has a software architecture in which both Data and Knowledge are separated from the computer program which manipulates them (which is called the Inference Engine).

3.1.4- KNOWLEDGE BASE ("KB")

It is the *model of the portion of the Universe which is addressed by an ES*. It may contain a description of all elements of the classes under consideration, or a list of their mutual relations, or a list of rules (including mathematical formulae) according to which to operate on these elements. It can contain procedures which point outside of itself, and create dynamic links to other procedures or data bases.

3.1.5- INFERENCE ENGINE (“IE”)

The software which uses the "data" represented in the KB, in order to reach a conclusion in particular cases. In general, it describes the strategy to be used in solving a specific problem, and it acts as a sort of "shell" which guides the ES from query to solution.

3.1.6- RULES

A rule is a way of formalising declarative knowledge: it consists of a Left Hand Side ("LHS") and a Right Hand Side ("RHS"), connected by certain sequences of logical symbols that make the sequence "LHS-connectives-RHS" equivalent to a "IF (LHS) THEN (RHS)" logical chain. The LHS is composed by a series of propositions to be regarded as "conditions", joined by AND or OR logical connectives; the RHS is a proposition which corresponds to an "action". The general meaning of a rule is: "RHS is TRUE IFF either all of the AND-joined conditions or at least one of the OR-joined conditions in the LHS assume a TRUE value". Example: "IF a machine is rotating AND IF it has blades AND IF a fluid expands in it, THEN the machine is a turbine". Notice that *a rule is a statement of relationship(s), and NOT a series of instructions*: the IF..THEN construct shown here is entirely different from the FORTRAN IF...THEN...ELSE construct. The latter is a *procedural conditional instruction*: if A is true, then *do* B, else *do* C; while a rule is *a logical relational statement between propositions*: if **p** is true and **r** is true ... and **s** is true then **z** is true. As a consequence of this definition, it is also clear that a rule can be also viewed as a *logical operator* acting on a premise (**p**) at a certain knowledge level and projecting the conclusions (**z**) onto a higher level: **z** becomes a meta-representation of the set (**p**, **r**, **s**,...).

3.1.7- FACTS

Facts are specific expressions which describe specific situations. These expressions can be numerical, logical, symbolic or probabilistic. Examples:

- "height = 130 m"
- "IF (outlet pressure) LESS THAN (inlet pressure)
  THEN (process is an expansion) IS TRUE"
- "p TRUE $\forall$ x $\in$ A"
- "IF (pressure high-frequency oscillations) is TRUE
  AND (shaft high-frequency vibrations) is TRUE
  THEN [probability(compressor stall) = 75%]
  AND IF (superimposed low-frequency pressure oscillation modulates high-frequency oscillation) is TRUE
  THEN [probability(compressor stall) = 95%]"

3.1.8- OBJECTS

An Object is *a computer structure which can contain data, data structures and related procedures*. It is best described as "a package of information and the description of its manipulation" [80]; another definition is “an Object is an entity that combines the properties of procedures and data, since it performs computations and saves local state” [74]. Example: "Francis hydraulic turbine with H = 300 m, Q = 20 m^3/s, whose performance is described in subroutine FRANCIS". Notice that the first portion of this Object is a declarative description, while the second one is a procedure (which could be, for instance, written in a conventional language like FORTRAN or C).

The traditional view of software systems is that they are composed of a collection of "data" (which represent some information) and a set of "procedures" (which manipulate the data): the code invokes a procedure and gives it some data to operate on. Thus, data and procedures are treated as completely distinct entities, while they are not: every procedure must make some assumption about the form of the data it manipulates. An Object represents both entities: it can be manipulated, like data, but it can describe a procedure as well, and manipulate other Objects (or itself). Objects communicate through messages, which are the manipulating instructions sent by one Object to the one(s) it wants to modify. A message contains the so-called "selector", which is the "name" of the desired manipulation: no details of the related procedure are sent with the message. This is an advantage, since procedures may be "local" to certain Objects, and differ (as it is usually the case) depending on the data they are applied to.

3.1.9- CLASSES

A description of one or more "similar" Objects. Example: "Hydraulic turbines". Notice that in this case, the "Object" described in 3.1.8 would be an *instance* of the Class introduced here. Since also Pelton and Kaplan turbines would, of course, belong to this Class, the corresponding Objects will bear a significant degree of similarity with the "Francis" Object, but will perforce have some different characteristic either in the description (e.g.: "Pelton hydraulic turbine") or/and in the data attributes (e.g.: "H = 1000 m and Q = 5 m^3/s"), or/and in the procedural part (e.g.: "whose performance is described in subroutine PELTON"). The definition of Class is recursive: there may well be sub-Classes (e.g.: "Francis", "Pelton" and "Kaplan") each one with sub-sub-classes (e.g.: "Francis with high specific speed", etc.), and so on. The limit to the subdivision is set by common sense: the closer the construct is to a real world entity, the easier will be the translation of the "real world problem" into a computer model vz. program.

The members of a Class (or of a sub-Class: the property is recursive by its own definition) are said to *inherit* all the properties common to and typical of the Class they belong to: naturally, they will have some distinguishing feature(s) which makes them different from other members of the same class, but they participate in the inference process with the properties they share with their fellow members. For example, sub-Classes "Kaplan", "Francis" and "Pelton", as well as "Steam Turbine" inherit the property "rotating wheel with blades" from their common ancestor-Class "Turbine".

3.1.10- INDUCTION

Sometimes erroneously called "abduction", is a logic procedure which proceeds *from effects backwards to their causes*. “Induction” is the method first proposed and formalised in 1620 by Sir Francis Bacon in his “*Novum Organum*”, and originally it meant the activity of deriving a scientific theory from a set of well-organised empirical data and some general rules of “abstraction” (thus subtly denying its own essence, because the principles of this “abstraction” must by force not derive from empirical observations!). In spite of some literary examples to the contrary (noticeably, Sherlock Holmes, who used almost exclusively inductive reasoning with an *amazing* degree of success), induction is far from being infallible, and has to be used with care, or together with some method for scanning all of the actually known (or foreseeable) causes an event may have, and for properly choosing among them, possibly by probability methods [20]. For example, high-frequency vibrations detected on a gas turbine shaft ("effect") may be caused by compressor stall, rotor unbalance in the compressor or in the turbine, fluid dynamic instabilities in the flow into the turbine, irregular combustion, mechanical misalignment of the shaft, failure in the lubrication system, bearings wear, or a crack in the gas turbine casing. Pure induction will not produce a solution in this case, because it cannot decide which one is the “exact” cause. Thus, additional data, in the form of secondary effects generated by each one of the causes, must be available which can be included in the inductive process to eliminate some of the causes from the list by matching the globality of the effects they would induce with the actually observed effects.

3.1.11- DEDUCTION

A logic procedure which proceeds *from causes forward to their effects*. Deduction is much more reliable than induction (of which is the logical opposite), but also much more "deterministically limited". For example, compressor stall can cause high-frequency vibrations on a gas turbine shaft; but if the *only* cause for gas turbine shaft vibration we have in our knowledge base is "failure of lubrication system", we will deduce (with a degree of certainty of 100%!) the wrong conclusion. So, pure deduction must be endowed of "double-checking" procedures which leave the possibility open for unexpected different causes to be present: in the case of the example, an oil temperature and pressure check could exclude the lubrication failure as a cause, and prompt for further analysis. With a perfectly symmetric procedure to that suggested above for induction, deductive inference must thus be supplemented by some form of empirical control of the cause-effect database.

3.1.12- BACKWARD CHAINING ("BC")

A procedure which attempts to validate or deny a proposition ("goal") by searching through a list of conditional rules and facts to see if they univocally determine the possibility of reaching the goal. BC procedures are also said to be "goal-driven". Example:

rule 1: *(Francis) IF (turbine) AND (hydraulic) AND (specific speed lower than 2 and higher than 0.3)*
rule 2: *(turbine) IF (rotating) AND (vaned)*
rule 3: *(hydraulic) IF (fluid = water)*
fact 1: *(rpm = 300)*
fact 2: *(volume flow rate = 30* m^3/s*)*
fact 3: *(total head = 500 m)*
fact 4: *(fluid = water)*
fact 5: *(number of blades = 13)*
fact 6: (*specific speed* $= rpm * \frac{(volumeflowrate)^{1/2}}{(g * totalhead)^{3/4}}$)

goal: "*given the knowledge base, determine if the machine is a Francis*"

(result: "*it is not*")

A backward-chaining algorithm methodically applies each rule in turn to scan the Knowledge Base to try to verify the goal. In this case, the "reasoning" would proceed as follows:

1. (rule 1 = TRUE) IF (rule 2 = TRUE) AND (rule 3 = TRUE) AND ($0.3 < n_s < 2$)
2. (rule 2 = TRUE) IF (fact 1 = TRUE) AND (fact 5 = TRUE)
3. (rule 3 = TRUE) IF (fact 4 = TRUE)
4. ($n_s < 0.3$) THEN (rule 1 = FALSE) THEN (THESIS = FALSE)

3.1.13- FORWARD CHAINING ("FC")

A procedure which attempts to *deduce* from a given Knowledge Base some "new" inferences. It is also said to be "data-driven". For example, working on the same KB as in 3.1.12, a FC algorithm would infer from Facts n. 1,2,3 and 4 that the specific speed is equal to 0.24; from Rule 2, Fact 2 and Fact 6, that the machine is a turbine, from Rule 3 and Fact 5 that it is hydraulic, and conclude from Rule 1 that it is NOT a Francis. The meta-code would look something like:

p = "IF (fact 1 = TRUE) THEN (machine is rotating)"
q = "IF (fact 4 = TRUE) THEN (machine is hydraulic)"
r = "IF (fact 5 = TRUE) THEN (machine is vaned)"
s = "IF (fact 6 = TRUE) AND (fact 1 = TRUE) AND (fact 2 = TRUE) AND (fact 3 = TRUE) THEN (specific speed = 0.24)"
t = "IF (p = TRUE) AND (q = TRUE) AND (r = TRUE) AND (s = TRUE)
THEN (rule 1 = FALSE) AND (rule 2 = TRUE) AND (rule 3 = TRUE)"
IF (t = TRUE) THEN (THESIS = FALSE)

This example clearly shows a fundamental characteristic of AI inference procedures: the fact that the machine is not a Francis is *already contained* in the KB, only, it is not explicit, and needs some "reasoning" to be extracted from it. The deductions of a FC algorithm are not "intuitive leaps": they are rather formally correct (albeit in some cases redundant, or trivial, or just plain uninteresting) conclusions: i.e., *effects* compatible with the given KB *causes*.

3.1.14- HIERARCHICAL REFINEMENT

The concept of Hierarchical Refinement consists in developing a design plan or implementing a process in descending levels of abstraction (see also Section 3.1.2: we are now proceeding in the *opposite* direction to the knowledge gathering process!). In essence, this is equivalent to the common Engineering practice of "leaving the details for the later stages of the design task". The refinement is always implemented from top to bottom ("top-down procedures"), and can be employed in Forward- and Backward chaining Inference Engines. Though conceptually straightforward, its implementation is complex and knowledge intensive. See also Section 5.1 for a somewhat more detailed description of the concept of "Meta-Level", and see [74] for a complete discussion on Hierarchical Refinement. One of the problems that arises when applying Hierarchical Refinement is that of *backtracking*, i.e., of recovering the (logical) point where the refinement can be resumed after a chosen path has failed. Since backtracking is essentially a problem connected with search methods, it is discussed with major detail in Section 3.11.

3.1.15- Blackboard Systems

During the solution of an Engineering Design problem, there may be instances in which a decision made at a certain stage about the (logical or numerical) value of a variable *x* bears a direct or indirect influence on another design decision already made at a previous stage about another variable *y*: in this case, it is necessary to "go back to the drawing board" and make a note that the value of *y* should be updated, and check that this modification is properly propagated throughout the design (notice that this problem has some analogy with the backtracking mentioned above). Another often encountered situation is one in which several patches of knowledge originating by different sources must be integrated into a single Inference Engine. Blackboard Systems are ideally suited to deal with these two situations. A Blackboard System consists of a number of knowledge sources (which can be formed by several physically distinct sources acting concurrently, as it would be the case in the second situation just depicted, or by different levels of the same source, as in the bactracking problem); each source communicates with a blackboard (though not directly with other sources), and the interaction is controlled by a specific Inference Engine (in principle, one for each source). The entire activity is in turn controlled by a "main" Inference Engine, which handles strategic decisions. The knowledge communicated from a source is checked and classified first by its own inference engine, which assigns an "executable" flag to specific chunks of this knowledge, and then by the main IE, which places them in the proper "slots" on the Blackboard. The Blackboard can be described as an agenda, or as an organised list, but probably the easiest way of thinking of it is that of a pseudo-logical matrix (we shall refer here to a two-dimensional matrix for simplicity), in which the lines represent the levels of the conceptual task to be performed, and the columns are the sub-levels, or the actions for each level. When a chunk of knowledge becomes executable, it is placed in the Blackboard in position (i,j), it is assigned a certain priority, and track is kept of all of its connections with other slots (r,s). If it turns out at a later stage of the inference process that another piece of knowledge belongs to that very same slot with a higher priority, then all the connections of the previous object occupying that slot are deleted, and substituted by the proper ones created by the new chunk of knowledge. The Blackboard (thence its name) is thus constantly "erased" and "rewritten" to reflect the most updated inference globally induced by the knowledge considered up to that point. The inference process is concluded when a scanning of all of the gathered knowledge does not change the status of the Blackboard (i.e., if there is no more "erasing" and "rewriting"): the final status of the Blackboard can now be executed, and it will result in the action cumulatively prompted by the set of knowledge chunks available. Figure 3.2 depicts the operational structure of a Blackboard System: notice that each "agent" is an Inference Engine, that there may be different Knowledge Bases (sources), and that a separate IE handles the net-to-blackboard transfer of information. By its nature, Blackboard Systems are well-suited for parallel (or concurrent) implementation: each agent could be working in principle by its own, on the nodal-CPU of a multi-node machine. Message-handling and other timing issues that trouble parallel numerical computations affect also this type of implementations: therefore, there are very few examples of successful, truly parallel codes.

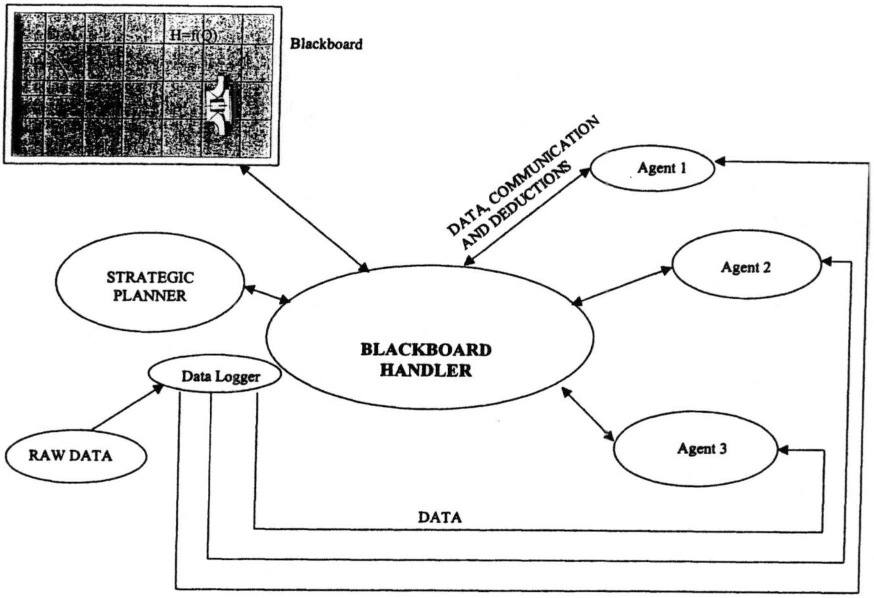

Figure 3-2. Operational structure of a Blackboard System

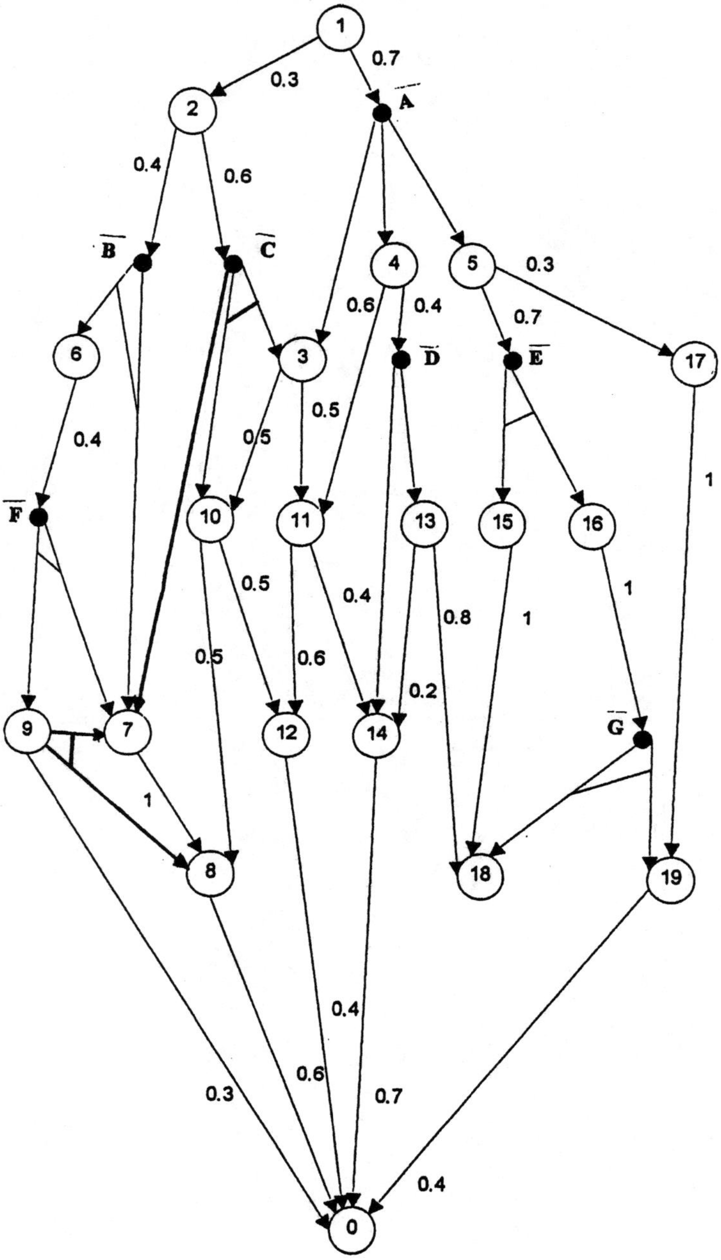

Figure 3-3. Tree representation of a hypothetical state space

3.1.16- DECISION TREE

A Decision Tree is a graph describing a multiple-choice situation (Fig 3.3). A given goal (the "root" of the tree, shown upside down in the figure) can be reached through the satisfaction of one or more *partial goals*, each one of which admits its own sub-goals, and so on. Since there can be (logical or physical) constraints on each one of the nodes (which represent the partial goals) or/and on each of the "branches" (which represent the *actions* which take from one goal to the next), in general *not all paths are acceptable:* the problem of finding a path from the root to the "leaves" (the basic goals, which can be viewed as the original facts representing the Knowledge Base) is a search problem (see Section 3.9). A "solution" is a path connecting the root with one or more leaves: it has to be remarked that the mere existence of a decision tree does not guarantee that a solution exists nor that, if existing, it can be found. Notice that the nodes need not represent homogeneous activities: in Fig. 3.3, the nodes numbered from 1 to 19 may denote sub-processes, while the nodes indicated by capital letters (A through G) signal main processes. Notice further that some of the branches are connected to each other by a fictitious line that clearly represents no physical process: this is a convention used to signify that the nodes from which the branches originate are "AND" nodes: so, the realisation of main process B requires *both* sub-processes 6 and 7; process C requires *either* 7 *or* (3 and 10). A node is said to be an AND-node if all the branches originating from it are AND-connected; conversely, it is said to be an OR-node if all the branches originating from it are mutually exclusive. In general, like in Fig. 3.16, most real processes are AND/OR, i.e., they are connected to branches of both modes (they are said to be *hybrid*: in the tree depicted in the figure, C and 9 are AND/OR nodes). Since it is logically convenient to reduce a tree to a set containing no hybrid nodes, but rather AND and OR nodes only, techniques for eliminating these mixed-mode nodes have been devised, and one is explained in the example below, Section 3.9.3.

3.1.17- KNOWLEDGE ENGINEER

The Knowledge Engineer is a person who is experienced in AI and ES and who may also have a background in Mathematics, Philosophy or Cognitive Psychology. His task is to understand the various ways in which knowledge can be structured and to create a framework for a KB so that it can be represented in a way which reflects the Domain Expert's way of looking at it, and can be accessed directly by both the Domain Expert and the ES user in a way (i.e., with a language and using procedures) which is familiar to both.

3.1.18- DOMAIN EXPERT

The Domain Expert is a person who is experienced in the subject matter of the ES and whose knowledge ought to be "infused" into the ES. This person may and may not be computer conversant, but he must have a good and complete understanding of the domain covered by the ES. Domain Experts need not be renowned academicians nor technological wizards: at times, their knowledge is at low level, while on occasion it may also be rather intuitive[3]. Notice that it is widely recognised that it is not its IQ, but rather the breadth or depth of his/her knowledge that make the expert: to quote from [17], “take an expert and place him in novel territory, and you get decidedly a non-expert behaviour. Conversely, take a genius and place him in the expert’s territory, and you do not get expert behaviour”.

3.2 - RELATIONAL VERSUS DETERMINISTIC PROGRAMMING

A sufficient condition to apply AI techniques to an engineering problem is *that there are no numerical methods immediately applicable to the solution of that problem*, either because a complete mathematical formalization of that problem does not exist, or because the resulting mathematical system of equations cannot be resolved in reasonable computer time on the available resources (see also [66]). There are of course substantial formal differences between the relation-based programming required by KB Systems and the deterministic programming employed by the numerical methods we are all accustomed to: but there is a subtler, yet more important philosophical difference, which we shall try to clarify here.

The action of "properly formulating" a problem, by posing it in the form of a system of equations with the appropriate boundary and initial conditions and with consistent constraints, can be conceived as a way of anticipately describing the solution at a certain abstraction level, by making use of an adequate symbolic language (e.g., the algebraic symbolism). Obviously, the quantitative (numerical) solution is not known a priori, because it will result from the calculations, but *the solution (and its very same existence!) is already contained in the descriptive equations*: in this respect, this deterministic approach is immediately directed towards the solution. In clearer terms, the formulation of an engineering problem in a form suitable for numerical solution rests upon the modellization of the underlying physical phenomena: and while it is not true that, once a model is formulated, its solution is already known, the solution is surely heavily model-

[3] *Intuition* is one of the cases (the other being *common sense*) for which we absolutely need an expert to gather some knowledge. It is indeed impossible to generate an AI computer procedure that displays genuine intuition. As for common sense, no program has yet been implemented that is capable of making the extremely wide logical leaps implicit in this apparently low-level human quality.

dependent. In a quite different fashion, in *knowledge-based solution methods*, there is no anticipate description of a solution: rather, this solution (if existing) must be sought after ("produced") on the basis of the knowledge of the *domain* in which a problem is posed (its *context*). In other words, the model itself must be - heuristically or logically or symbolically- constructed by the solution paradigm. Numerical methods result therefore in a system of equations whose *form* implicitly represent the general features of the solution (features which can in fact be estimated by studying the "asymptotic" behaviour of the given equations); knowledge-based methods result instead in a formulation which describes the *context* in which a problem arises. And, while numerical methods require numerical techniques which solve the governing equations *with a degree of accuracy established by the user and in principle reducible at will*, knowledge-based methods make use of the rules of predicate calculus to infer (deductively or inductively or probabilistically) the solution from the available information, and the problem position consists of both the information and the rules.

A first difficulty is thus that of defining a representation of the "problem context" which also allows the representation of inference rules: each rule must be by necessity defined as a (logical) function of the knowledge it represents, to which it can be applied and which it, in some way, manipulates. As a consequence, and only in this sense, the problem takes on a double valence: the procedure and the data on which it is (or may be) called to operate *must be represented with the same set of symbols*.

3.3 - LOGICAL SYSTEMS

An ordered and "complete" set of correlated symbols which can be used to represent the general knowledge and the specific or generic facts from which a solution to a certain problem can be found, is the language of mathematical logic, and in particular its sub-set of propositional and predicate calculus [32,35,37]. Notice that Propositional Calculus consists of logical propositions joined by logical connectives (AND "$\cap$", OR "$\cup$", NOT "$\neg$", IF...THEN "$\Rightarrow$", and EQUAL TO "$\equiv$"), and cannot handle either quantifications ("p LARGER THAN q") or variable binding ("FOR ALL q $\in$ X, p(X) is TRUE"). Predicate Calculus allows for both, and is therefore used in quantitative decision making procedures, while Propositional Calculus is best applied in pure inference. Note that when we refer to "Predicate Calculus" in this book, we shall always mean "First Order Predicate Calculus", i.e., the form in which the variables are allowed to be assigned quantitative values, but the functions (=the predicates) are not.

A logical system is defined as a coherent system consisting of *signs* and *rules* for the correct use of these signs. As such, it is not a *theory* proper, because it is not limited to a set of sentences about a specific object or type: it is in fact entirely equivalent to a *language*. For what is of interest here, we can identify two main factors in the study of a language: its *syntax* and its *semantic*[4]. The syntax of a language establishes the rules for constructing and transforming it: it defines which signs are to be used in a given context, how these signs can be connected to each other, and how they are to be manipulated. Semantic is concerned with the interpretation of the signs the language consists of, and decides about the *values* to be assigned to each sign or about *interpretation rules* for each sign or group of signs in a given context. For the language to be (in some sense) *meaningful* it is necessary that the individual signs or ordered sets of signs can -and in effect do so- represent our knowledge in some application domain. Furthermore, the sign evaluation rules must guarantee that the transformation ("translation") procedures be evaluation-invariant, i.e., that they preserve the value of sign concatenations.[5]

Once a language has been proven to be "complete" and meaningful (see also [55]) it follows that from the application of any combination of transformation rules to a "true" proposition, another "true" proposition results. Notice that, in this context, "true" means "logically true", and can, in fact, be "physically wrong". To avoid confusion, we will use "false" and "wrong" as the opposite of "true" and "correct" respectively. The difference can be clearly represented by the following examples:

Example 1:

Rule 1 (verbal expression): "The efficiency of two components connected in series, such that the only useful output of the first is the only useful input of the second one, is equal to the product of the two component efficiencies"

This rule corresponds to a meta-code like:

Rule 1 (meta-code formulation):
IF (C2 ⟨in-series-with⟩ C1) AND
p = "η_1 = (efficiency of C1) AND η_2 = (efficiency of C2)" AND
q = "η_{tot} = (efficiency of C1+C2)" AND
p = TRUE AND q = TRUE
THEN
r = "$\eta_{tot} = \eta_1 * \eta_2$" is TRUE

[4] The *grammar* of the language is of course of paramount importance for its correct understanding, but we shall assume here that it can be always reconduced to a complete and coherent set of connectives and operators applied to the symbols.

[5] Actually, there is much more to be said about the concept of language (see [55] for a yet unsurpassed reflection on the meaning and limits of a language). It can be remarked here though that the study of logical systems (i.e., the theory of language) leads to a better definition of the possible and permissible extent of our Knowledge Domain: in his latest work Wittgenstein [81] concluded that "the variable '*proposition*' must be of the same order of [meaning: "must possess the same attributes as", note of these Authors] the concepts of '*Reality*' or '*Universe*'".

This is an example of a logically true rule. However, if wrong values are assigned to η_1 and η_2, application of the rule can result in a physically wrong (though logically true!) statement:

IF ("$\eta_1 = 1.3$" AND "$\eta_2 = 2$") THEN

Rule 1 $\langle C1, C2 \rangle \Rightarrow$ "$\eta_{tot} = 2.6$")

To avoid this, additional rules ought to be inserted, that ensure for instance that $0<\eta_1\leq 1$, $0<\eta_2\leq 1$.

Example 2:

Suppose the prescribed task is the generation of a schematic layout of a process to produce P_{el} MW of electrical power (see Example 6.6). The "conventional" option (which we shall refer to as the "A" configuration in the following) is to install an electrical generator whose output is P_{el}: this generator will require a mechanical input $P_{shaft} = P_{el} / \eta_{gen}$. To produce P_{shaft}, the first four options which come to mind are:

B) a Diesel engine, with a fuel input of $m_{fuel,Diesel} = \dfrac{P_{shaft}}{\eta_{mech}\eta_{diesel}LHV_{fuel,Diesel}}$

C) a turbogas set, which requires a fuel input of $m_{fuel,TG} = \dfrac{P_{shaft}}{\eta_{mech}\eta_{TG}LHV_{fuel,TG}}$

D) a steam plant, which requires a fuel input of, $m_{fuel,ST} = \dfrac{P_{shaft} + P_p}{\eta_{mech}\eta_{steam}LHV_{fuel,steam}}$ if P_p is the power absorbed by the condenser extraction- and feedwater pumps.

E) an electrical motor, which requires as input an amount $P_{motor} = \dfrac{P_{shaft}}{\eta_{mech}\eta_{motor}}$ of electrical power.

So, the following "energy conversion chains" are *all logically true*:

1) $\langle B+A \rangle$ = fuel chemical energy→Diesel engine→electrical generator
   Result: P_{el} kW of electrical power are generated by burning
   $P_{el}/(\eta_{gen}\ \eta_{Diesel}\ \eta_{mech}\ LHV_{Diesel})$ kg/s of Diesel fuel;
2) $\langle C+A \rangle$ = fuel chemical energy→turbogas set→electrical generator
   Result: P_{el} kW of electrical power are generated by burning
   $P_{el}/(\eta_{gen}\ \eta_{TG}\ \eta_{mech}\ LHV_{TG})$ kg/s of light oil or natural
   gas;
3) $\langle D+A \rangle$ = fuel chemical energy→steam plant and turbine→electrical generator
   Result: P_{el} kW of electrical power are generated by burning
   $P_{el}/(\eta_{gen}\ \eta_{steam}\ \eta_{mech}\ LHV_{steam})$ kg/s of heavy oil, coal or natural gas;
4) $\langle E+A \rangle$ = electrical power→electrical motor→electrical generator

Result: P_{el} kW of electrical power are generated by tapping
$P_{el}/(\eta_{gen}\ \eta_{motor}\eta_{mech})$ kW of electrical power from the net!

While it is clear that alternative 4 is *wrong*, it is not *false*: we shall see below (Example 6.6) that to instruct an ES Process Designer to avoid this choice we must *make it false*, and negate its logical possibility by adding a rule like:

IF (P_{el} = product) AND
$P_{auxilaries}>0$ AND
($P_{network}$ = power tapped from the net) AND
$P_{auxiliaries} \leq \alpha P_{el}$ AND
$\alpha << 1$
THEN
"$P_{auxiliaries} = P_{network}$" is TRUE
where the factor α (small but not equal to zero) takes into account the possibility of electrical pump drives, or other similar auxiliary drives, but excludes the occurrence of alternative 4.

Both Propositional and Predicate Calculus have been recognised to be "complete" and meaningful, and enjoy a higher degree of simplicity and efficiency ("power") than other languages; they are therefore our preferred tools for AI applications[6].

The representation of our knowledge of a specific application domain in a Logical System, i.e., the projection of our perception of the facts and rules belonging to the context of that problem onto the symbol-space pertaining to that system, constitutes the basis upon which the Logical System draws its subsequent inferences, to establish "new" facts and "new" rules ("new" being used here to significate "not explicitly contained in the original formulation of the problem").

The mechanisms by which inference can be gained are called, in AI terminology, *inferential* (or "*inference*") *engines*: they consist of a set of procedures defined in one of the logical languages, which can derive from a given batch of knowledge some information which was not explicitly contained in that knowledge. They do so, *independently of the physical "value" of the initially imparted knowledge*, i.e., they are indifferent to the "correctness" or "wrongness" of the Knowledge Base. There are basically two types of inferential

---

[6] A related topic, which will only be mentioned here, is the co-ordinated series of interdisciplinary attempts aimed at the realisation of a complete and reliable method capable of translating a natural language into a logical one: though the problem is far from being even completely formulated, its solution would have a significative impact on AI techniques, because natural language is the most natural form of expressing our own knowledge, and since a fully automated "translator" could be conceivably built into a computer code, the possibility would then exist of building *an electronic "inference engine" which would not use intuition but exact logical reasoning to draw correct conclusions from the description of the world we would impart to it via our natural language.* This futuristic scenario is -in reality- much farther away than specialists thought it to be only a few years ago [20, 60].

engines: the backward-chaining ("BC") and the forward-chaining ("FC") engine. A BC-engine (Section 3.1.12) tries to verify whether a fact or a rule (called "goal") can be deduced from the Knowledge Base, and is, in this sense, said to be "goal-driven". The search proceeds backwards from the goal to a set of facts or rules which could be its cause and either are in the KB or are deducible from it. The method works recursively, until either a reverse causal path has been established or the path is broken and the search has failed. To properly apply the BC-engine, it is of paramount importance to be able to impart to it a precise, complete and detailed description of the goal in all of its attributes. A FC-engine (Section 3.1.13) consists of a set of procedures which try to infer from the facts and the rules contained in the knowledge base (which it regards as "prime causes") some logically acceptable effects. This method, too, works recursively, so that a chain of primary, secondary, tertiary,.... effects can be built. The FC-engine is said to be "data-driven". It has to be remarked at this point that the final effect inferred by a FC-engine, being at a different abstraction level than the objects forming the knowledge base, may appear to be strikingly different from them, and give the impression of a "intuitive leap" in the inference process: indeed, it is not so, and the causal chain could always, at least in principle, be deduced "manually" with standard logical operations.

Some of the most recent applications combine the two approaches in an "Alternate Direction Search" ("ADS"), in which one branch of the search *deduces effects* from the knowledge base (BC-engine), while the other branch tries to *induce causes* from the goal (FC-engine): an ADS "solution" is reached when the last effect from the FC-engine coincides with the first cause of the BC-engine.

Once completely formulated and implemented, both a BC- or FC engine result in procedures which are not capable of handling *a change in the universe of application* (a typical example of which is a problem in which the solution depends on the technological level of the environment, which of course is changing with time). This rigidity can be overcome to a certain degree by a careful initial formulation of the original Inference Engine, but it is in general impossible to foresee all of the changes which can affect the domain of application of an ES.

An approach which can be used to solve this problem is the *blackboard approach* described in Section 3.1.15, which will be described here from a slightly different viewpoint. Knowledge is subdivided into sub-domains S_i, on each one of which a separate inference engine IE_i operates. The entire problem is formulated as a succession of partial solutions which may be at different levels of abstraction. These partial solutions P_i constitute the "blackboard" and there is a main procedure M which calls in a certain sequence those IE_i's which -in its judgement- can produce a modification to the overall state of the blackboard: each of these IE's, operating on its own Knowledge Base, produces some changes in the current blackboard state. In turn, other IE's examine the state just updated and decide whether they too can contribute to a further updating. The whole process is controlled by the central inference engine M. The process is repeated until either there is no modification in the blackboard state and the goal has not been achieved (failure) or the global goal has been achieved, disregarding whether the blackboard state can be further modified or not. It is apparent at first glance that this description of a Blackboard System bears a strict resemblance (a logical analogy) with the logical scheme used by interdisciplinary design teams: they (try to) put together different levels and types of knowledge by means of a hierarchically structured procedure.

In spite of the formal completeness and mathematical rigor of Logical (symbolic) Systems, human thinking activities have been shown to *exceed* their representation capabilities: i.e., there are "anomalous" knowledge descriptions which, though perfectly acceptable to a human expert, cannot be formalised (and thus, cannot be manipulated) by any Logical System. The most important examples of such descriptions are:

1) vague or approximate descriptions (example: "if the evaporation pressure is *too high* and there is no reheat, then there *might be* condensation *somewhere* in the final stages of the steam turbine")
2) hidden assumptions (example: "for a given steam throttle temperature, optimise throttle pressure"; here, the hidden assumption is "the process is not supercritical")

Vague descriptions can be tackled with a technique based on the so-called "Fuzzy Sets Theory", which is a part of Logical Sets Theory and which is discussed in Ref. [46]. The way to formalise approximate reasoning is briefly described in Sect. 3.8. Hidden assumptions are more difficult to handle, because they are invariably case-dependent. Simple, general cases can be treated using non-monotonic logical systems (see [20]: their treatment is also beyond the scope of this Book). However, for what concerns engineering applications, it is probably advisable to take a more pragmatic approach, and try to embed in the Knowledge Base the highest possible degree of flexibility, so that -for instance- there may be almost-redundant descriptions of the same fact which can still be processed by the Inferential Engine. It is auspicable that, in a not too far future, rigorous and formally exact omnicomprehensive logical treatments of anomalous thinking processes may be developed, but in the meantime simpler, more down-to-earth and robust applications ought to be implemented using our present day AI techniques.

3.4 - SEMANTIC NETWORKS

A Semantic Network consists of a set of nodes and of their connections: each node represents a *single piece of information*, and it can be a noun, an individual of a certain class, an activity, a problem, an event, etc.. There are no explicit limitations on *what* a node can represent. The intranodal connections represent the relationships between the nodes, and they cannot be arbitrary, but *are chosen from a pre-defined set of logically and physically possible relationships between the objects represented by the nodes.* In this sense, there is an implicit limitation to the information a node can represent: no node can be included in the network if there is no relationship available which fits it. This is only a formal problem though, because it is always possible to define an *ad hoc* relationship for any "special" or "anomalous" event.

Any triplet "node A-connection-node B" represents in a Semantic Network one of the possible states of its universe: given the extreme flexibility of the structure, it is possible to move on from a triplet to another one, both in the direction of higher complexity or towards higher simplicity. The most important characteristic of this type of knowledge representation is though that of being absolutely neutral with respect of any goal or direction: they can (be used to) respond to very different (even conflicting) requests, because the only "facts" they represent are formal relationships between states.

For this reason, semantic networks have enjoyed a remarkable degree of success in AI applications, and have even resulted in particular macro-structures, called *Object-Oriented languages*, and to some very powerful paradigms, collectively constituting *Object-Oriented programming* [74]. See also Section 4.3.

3.5- Fuzzy Sets

The rule-based approach employed by AI methods consists essentially in solving a problem by describing the relationship between causes (input data, "facts") and effects (output data, "results"). This approach implicitly subsumes a binary logic: either a fact is true or it is false, and consequently, either it produces the effects implied by the applicable rules or it does not. It must be recognised though that a large part of the engineering expertise encapsulated in the Knowledge Base possesses heuristic characteristics directly derived by subjective judgements, and as such, *it is frequently expressed as incomplete and imprecise information*. Application of classical logic to such an uncertain KB is likely to produce imprecise and incorrect results. Methods have been proposed to deal with uncertainty: Qualitative Reasoning (which requires a specific type of extension to the logical connectives, see Section 3.8), Probability Methods (which produce of course probabilistic results, see [74] for an extensive treatment) and a rather unique theory capable of directly dealing with uncertainty, called Fuzzy Sets Theory ("FST"). Fuzzy Sets Theory was introduced in the mid 60'es as a tool for dealing with the so-called "uncertainty of type 2", i.e., with the special, non-probabilistic type of uncertainty arising from human thinking patterns and from reasoning and perception processes, and in general from cognitive information (by contrast, "uncertainty of type 1" deals with the random-noise affected information arising from unfiltered measurements taken on physical systems). As an example, when monitoring a complex system or a process, it is customary to regard its behaviour as a continuous flow of information that can be decomposed into discrete events (the elementary facts) and subsequently explained in terms of the propagation of these events in the environment (context). But, this propagation of events is a very complex phenomenon which is usually modelled by the Domain Expert in the form of approximately correlated events, and of logical connectives which in turn assume an approximate rather than a precise meaning. In effect, to cope with the complexity of the diagnostic phenomenology, the expert builds a mental model based on previous experience, which qualitatively reproduces the structure of the correlations of events as a soft network consisting both of approximate links and approximate nodes. In addition, one has to consider that the performance of continuous dynamic systems is almost invariably characterised by a gradual transition from the range of "acceptable" to that of "unacceptable" behaviour: a sharp boundary between faulty and non-faulty operation is seldom given.

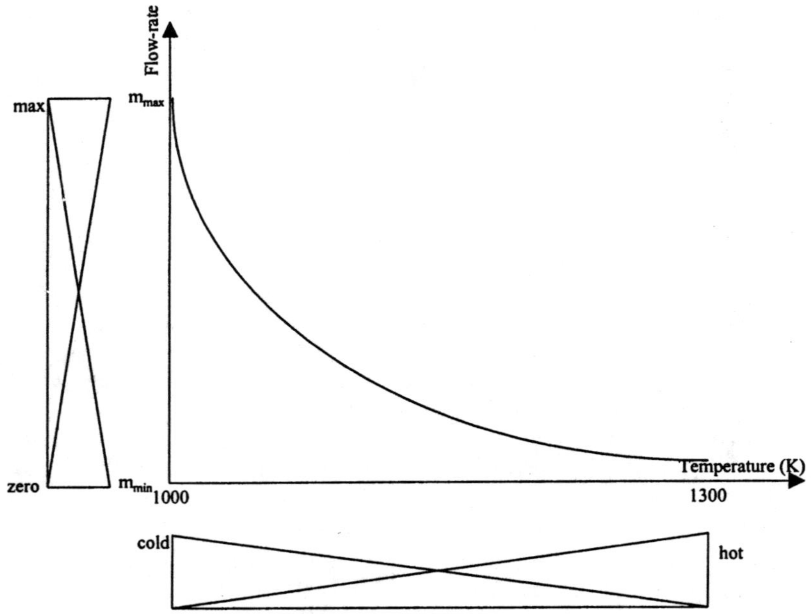

Figure 3-4. Fuzzy representation of the temperature-mass flow rate relationship in an FBC

There is therefore a need for a tool which can make use of the "approximate" database to infer conclusions according to inference laws valid for approximate reasoning: the laws of adjunction, of simplification, of *modus pollens* and the law of addition in the classic binary propositional logic are of course still valid, but their inference power is by force limited by the very nature of the available KB. What is needed are extensions of the type "if more than an assigned percentage of the conditions in the antecedent of a rule is satisfied then the consequence is true". This clearly represents a great enhancement in the ability of an Expert System to capture all of the knowledge embedded, explicitly and implicitly, in the KB.

Fuzzy sets theory ("FST") was first developed for process control applications, in cases in which the applicable rule(s) had an approximate expression, e.g.: "if the fluidized bed temperature is *too high* and its time derivative is *not too steep*, then the operator should *slowly reduce* the fuel mass flow rate, and throttle the fluidizing air *proportionally to it*". This is an example of a procedure which would be perfectly intelligible to a human operator, but which is extremely difficult to translate into computer instructions.

In normal Set Theory (pertaining to classical, binary logic), an object **O** either *is* or it *is not* a member of a certain class **C**: if we express this in mathematical terms, the *degree of belonging* (or *degree of membership*) of **O** to **C** is either 0 or 1 (in the above example, either the temperature is above 1300 degrees K or it is not). In FST, the degree of membership can take non-integer values between 0 and 1: again with reference to the above example, there will be a range of temperatures for which "too high" is *definitely true* (degree of membership of the object "temperature" to the class "too high temperatures" is 1, see Fig. 3.4); a range in which it is *definitely false* (degree of membership of the object "temperature" to the class "too high temperature" is 0), and an intermediate range in which it *can be true*, with a degree of "truth" larger than 0 but less than 1. These different ranges are called the *influence domains* of the degree of belonging, and are the basis for the development of a Fuzzy Set solution, which is obtained in three steps:

1. *quantification*: a rule is formulated, which attaches a certain "consequence" to each of the possible influence domains. This consequence (in the context of the above example, the amount of the corrective action) can be itself fuzzy, i.e., it can assume values like "small", "normal", "large" and similar. Expert Systems are based on the validation of the statements in the antecedent side of a rule (the *premises*). As recalled above, the general form of a rule is "IF **p** AND **q** AND **r** AND ...THEN **s**": now, for all cases in which the antecedent is the combination of at least two statements (**p** AND **q**), since in classical (binary) logic **p** and **q** must be either true or false, the statement (**p** AND **q**) is true only if both **p** and **q** are true. How do we define the fuzzy truth value of the statement (**p** AND **q**) on the basis of the separately known fuzzy truth values of **p** and **q**? The answer is: in Fuzzy Logic the truth of (**p** AND **q**) is the *minimum* of the truth values of **p** and **q**, taken separately.

   Similarly, consider the statement (**p** OR **q**), which in classical logic is true if either **p** or **q** or both are true, and is false only if both **p** and **q** are false. In Fuzzy Logic the truth value of (**p** OR **q**) is the *maximum* of the truth values of **p** and of **q**, taken separately. We see that FST is basically a *multivalued logic* that allows for intermediate values to be defined between conventional binary value assignments like yes/no, true/false, black/white, 0/1. Notions like "rather warm" or "very cold" can be formulated mathematically and processed by computers. Fuzzy Logic offers a more complete way of representing reality: in Fuzzy Logic a statement is true to various degrees, with a continuous gradation from completely true to completely false (confidence scale from 1 to 0). FST thus gives us a whole new approach to the mathematics of thinking: it is a change of paradigm with respect to the two-valued logic. For instance, in FST it is very likely that two or more descriptions might have non-zero confidence values at the same time, which is acceptable since in the real world (or better said, in our model of it!) it is very seldom the case that one description is entirely right and all of the others entirely wrong.

2. *rule deduction*: the (fuzzy) variable to be used as a control variable is measured: if its value falls into an influence region for which the degree of belonging is 0 or 1, then *a value 1 is assigned as a weight of the corresponding rule; otherwise, the non-integer value of the degree of belonging is used as a weight.* If more than one variable contribute to a rule, the *degree of executability* of that rule is given by $d_i = min(d_{ij})$, where i is the index of that rule and $j = 1...n$ is the index of the variables of interest. Notice that there can be cases in which the solution is already in integer form: if $d_i = 0$, the rule does not apply (its consequence will be the empty set, i.e., no action is required); if $d_i = 1$, the rule *certainly* applies (i.e., the full corrective action is required). If d_i is between 0 and 1, a further step is necessary to obtain the solution;
3. *defuzzification*: this step consists in *attributing an exact value to the degree of applicability of a rule for which d_i is between 0 and 1.* In our simple example, the most reasonable solution would be that of *assigning an "intensity" d_i to the consequence implied by the applicable rule*, i.e.: "decrease the fuel mass flow rate at a rate d_i/D_{max}, if D_{max} is the maximum fuel throttling rate allowed in operation. Generally, there will be more controlling variables and several composite rules (in the sense that each rule i will share some of the n variables with some other rule): in these cases, there are several possible solutions to the problem of defuzzification: one can take as intensity of the consequence an integral average of the d_i's, or its median value, or its most probable value, etc. There is no *a priori* valid general rule, and a solution must be heuristically found from case to case.

There may be cases in which the input data (the available knowledge bits) are in numerical form: then, we need a way to translate the input numbers into confidence values expressed in propositional form. This is done by applying a procedure known as *fuzzification*, which is a way of decomposing a system input (and possibly its output) into one or more qualitative fuzzy sets. This is achieved by using the so-called *membership functions*, that are in essence curves mapping the numerical data onto a set of values of the degree of membership of the relevant parameter.

Use of Fuzzy Logic is appropriate in very complex processes, where there is no simple mathematical model, or where the process is highly non-linear, or when the problem appears to have no known mathematical solution. By contrast, Fuzzy Logic is inappropriate in conventional control of processes where a mathematical model of the process is available.

Some examples of FST Logic applications in the field of Thermal Systems can be found in automatic control of dam gates for hydroelectric power plants; in air temperature control of air conditioning systems; in internal combustion engines control; in real-time control of high-efficiency industrial processes; in the control of glass melting furnace operation, etc.

3.6 NEURAL NETWORKS

"Classical" AI, as described in the remaining Sections of this Chapter, is based on the notion of *cognitive symbolism*, i.e., on the assumption that knowledge can be subdivided into "logical information bits" which can individually be represented by some proper symbol, in the frame of a pseudo-language which enjoys all the properties of our common language but does not have its shortcomings (double meanings, linguistical paradoxes, homophonies, referential opacities, etc.). Recently [38, 39], a different approach has been proposed for knowledge representation and manipulation, which is based on an entirely different concept: this approach is called *Cognitive Connectionism*, and considers knowledge as being distributed across a structured system (possibly, but not necessarily, linguistic or symbolic), so that *it becomes impossible to associate a particular portion of knowledge to any single parts of the system.* Since Cognitive Connectionism was first developed within the framework of a functional model for the representation of human brain activities, the AI methods originating from it collectively go under the name of *neural networks*, which is a reminder of their "neurological" origin. Neural Networks can therefore be studied both as "models" of the human brain and as methods for solving cognitive problems. We will consider them here only under this second point of view, comparing them -in perspective- not only to traditional logic-symbolic systems, but also to classical numerical methods.

Biological systems perform pattern recognition, which is an intrinsically very complex activity, by executing several computations carried out by brain neural cells: this is a well-known and well-established fact in biological- and cognitive science. Even before the emergence of Artificial Intelligence, researchers were attempting to reproduce the same mechanism in analogical or discrete electronic devices: when the capabilities of digital computers developed to an extent that they could be considered reliable tools for this purpose, AI researchers started trying to construct numerical models of this highly qualitative task. A series of confirmed observations has established that, though the single neurone computing capacity is relatively slow (a few milliseconds), the overall computation is carried out in few hundred milliseconds: this suggests that several computational steps are executed in parallel by a multitude of cells.

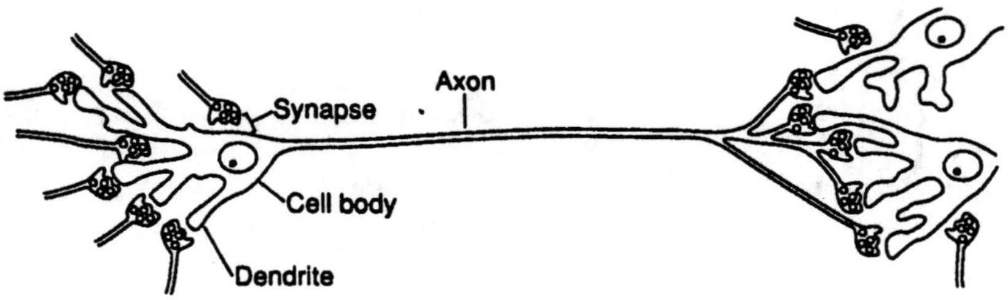

Figure 3-5. Simplified diagram of a neurone

In the late 1950s Frank Rosenblatt carried out experiments on neurological mechanisms using the "Perceptron" (a semi-analogical device based on electronic valves and switches); almost forty years later, in 1986, David R. Rumelhart and James L. McLelland published the book "*Parallel Distributed Processing*" [38], which fired a new interest and highlighted new applications of the Artificial Neural Systems (ANS) beyond the understanding of cognitive processes, including engineering applications, in which there is the need to adapt system behaviour to new situations on the basis of experience gained by the analysis of past similar situations. The question is: can a computer *learn* from past experience and use the newly acquired knowledge to derive solutions to new and ill-defined problems in classification, diagnosis, forecasting and analysis? This question actually implies a new question: What constitutes a learning process in a biological individual? We can assume that "learning" is composed by two sub-processes: acquisition and elaboration of data sets and storage of data sets into memory areas. To gain a better understanding of the complex phenomena that are involved, let us examine in some detail the biological neural cell system. Consider the structure of a neural cell as depicted in Fig. 3.5: a neurone consists of a *body*, with a *nucleus*, and a number of extensions called *dendrites* at the end of an a prolonged branch protruding out from the body (which is called the *axon* of the neurone). Branches at the end of the axon form *synapses* (connections) with dendrites of other cells. Signals are exchanged through synapses only when the signal intensity at the origin exceeds a certain threshold value. Synapses may perform an *excitatory* function, if the signal is added to (in a sense that it will become clear in the following) the neurone cell or an *inhibitory* one, if the signal is subtracted from the cell. Based on cognitive and biological science studies published as long as 50 years ago, a theory of how neural activities combine to produce what we call intelligent response to an external stimulus has been elaborated [39]. This theory is based on the following assumptions:

- a neurone can be regarded as a binary switch, its only two possible states being *on* ("excited", sends a signal) or *off* ("unexcited", sends no signal)
- a number of synapses (>1) must be excited in a certain period of time for a neurone to be excited
- the stimulus-induced signal must be above a certain minimum level, below which it cannot be perceived (it is "filtered out")
- the stimulus-induced signal must last for a certain minimum time, below which it cannot be perceived (it is "removed")
- the inhibitory activity prevents excitation of the neurone
- though the structure of the interconnected cells does not change, the structure of their interconnection may change as a result of external stimuli.

It is known that neural cells do not possess all their "knowledge" and "deductive capabilities" as inherited genetic qualities: they go through a continuous learning process. Suppose for instance that multiple signals concur to "fire" an action on a receiving neurone. If the excitation process is repeated persistently the receiving neural cell will "learn" and it will "fire" whenever it receives a signal from *any single one* of the exciting neurones[7]. This reflex can be extended and generalised, so that we can answer the first part of the initial question. The first phenomenological component of the "learning" process, namely, data acquisition and elaboration, is performed by the following scheme:

1. some primary stimulus reach an "encoding" area of the brain, where they are selectively sent to certain groups of neurones under the form of excitatory/inhibitory signals;
2. these first group of neurones responds to the secondary stimuli (the excitation signals) by "firing" or "not-firing" their own response to a second group of neurones they are permanently connected with, in a way that is totally (and deterministically) established by their present state and by the number and intensity of the signals they just received;
3. a second, third... group of neurones repeats the process: after a certain number of steps, a final *n-th* group of neural cells is reached that reacts to the *(n-1)-th* stimulus by firing what appears to be the brain's response to the original stimulus.

[7] This is the so-called conditioned reflex, and it is a very favourable evolutionary trait (for defense, or escape)

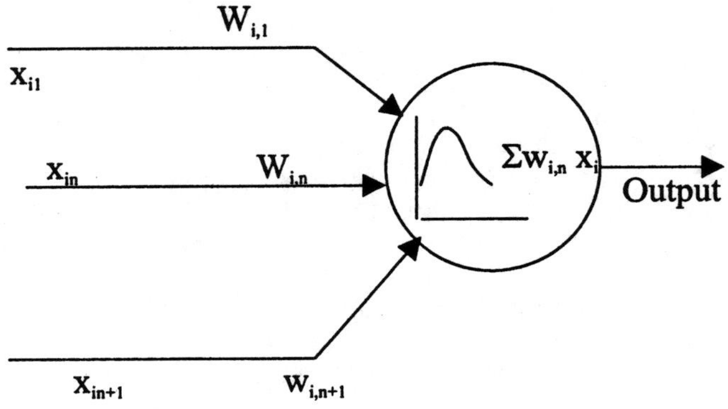

Figure 3-6. Artificial neuron activity and its output function

The answer to the second part of the question is not so simple, and neither there is univocal agreement among cognitive scientists: it seems that "memory" of past experience is stored in a Biological Neural System by means of at least three different mechanisms.

1. after the final response from the *n-th* layer is fired, the resulting action is scrutinised by some other portion of the brain, and its results "translated" in an adjustment of the response capability of the neurones involved in the chain that led to that action. This adjustment is implemented either in the form of a modified "inertia" of a neurone to respond to the same stimulus, or by a variation of its output in response to an equal level of stimulus. We say that in the first case the neurone's firing threshold has been increased (or lowered), and in the second the activity of the neurone has been modified. This is equivalent to remember what the reaction to the original stimulus was, to evaluate its desirability, and to modify the response by somehow "hard-coding" into the Neural System a corrective action;
2. it seems that for certain stimuli (typically, those connected with the most qualitative cognitive actions, like reconnaissance of similarities or detection of analogies, etc.), the *chain of causal relationships* can be stored (possibly in specialised neuronal structures) in such a way that if it is recognised that the first stimulus is approximately equal to one of which there is previous experience, then the chain of relationships is directly applied to it, and the same result (final action on the part of the *n-th* group of neural cells involved) is immediately produced;
3. finally, there is evidence that the connectivity of neural structures can be modified in an evolutionary sense: if the brain somehow "detects" that a portion of the reaction of its own neural structure to a certain stimulus is either useless or insufficient, it can modify the neural connections of that area, by inactivating some of the synapses and activating new ones, and then

"optimise" the response of the resulting new structure on the basis of "memory" of past causal relationships.

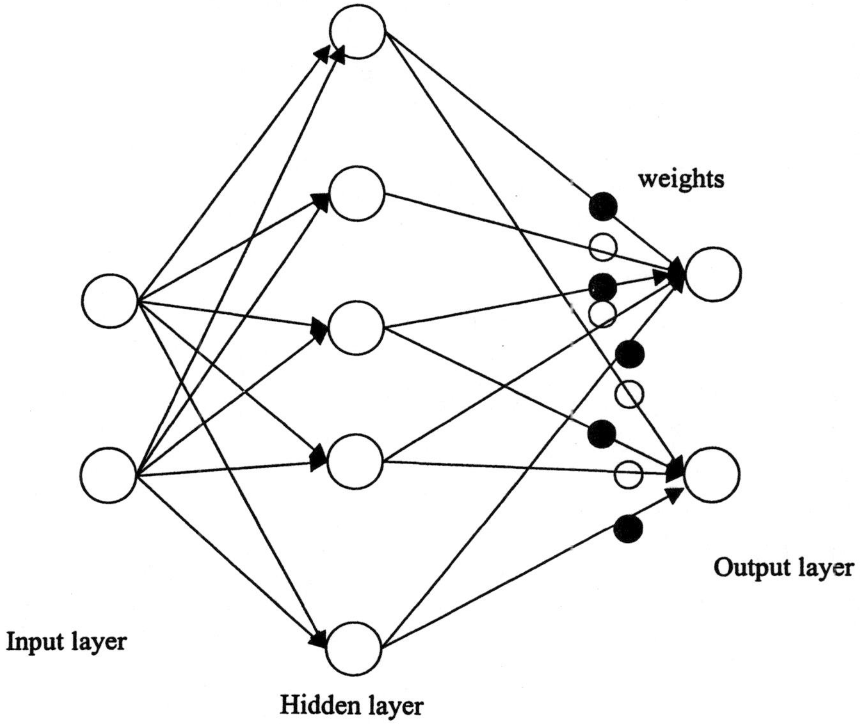

Figure 3-7a. Structure of single layer, feedforward Neural Network

The transposition of neural cells activity into Artificial Neural Systems (usually denominated "Artificial Neural Networks" and indicated by "ANN" in the following) is made possible by realising an interconnected network of ordered "processing elements" (or simply "processors"). Each processor's activity can be represented as in Fig 3.6. There are different neural network structures:

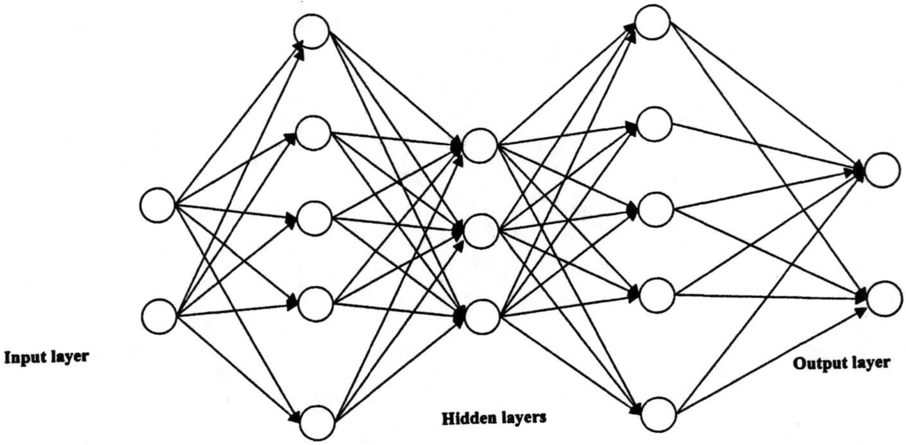

Figure 3-7b. Structure of multiple layer, feedforward Neural Network (weights not shown)

- *Pattern association*: this configuration is illustrated by the feedforward network illustrated in Fig. 3.7
- *Associative memory*: derived by the Hopfield model shown in Fig. 3.8
- *Self-organising networks*: also called unsupervised learning networks, their main characteristic being the capability of organising input patterns into classes on the basis of some common features.

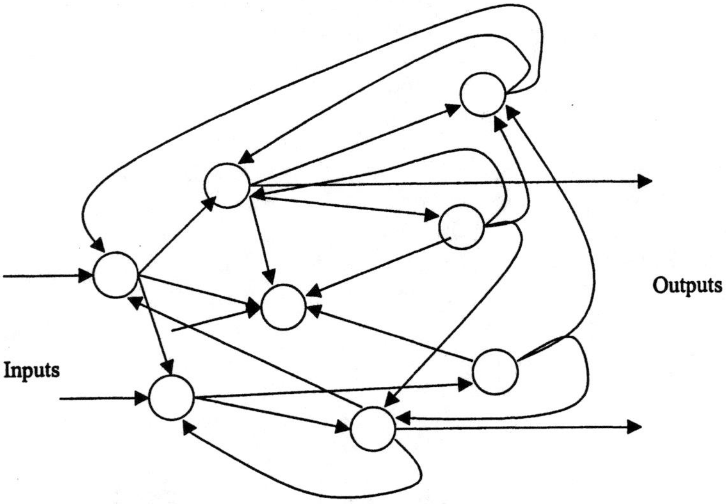

Figure 3-8. Hopfield model Neural Network structure

A Neural Network is thus a set of nodes (the *neurones*) organised in *layers*: each layer consists of a certain (possibly very high) number of neurones *not interconnected to each other*. But, *each neurone of each layer is connected to all neurones of the "upper" (the preceding) and "lower" (the following) layer*. The first layer is called the input-, the last one, the output layer. The connections between neurones are called, by analogy with neurology, *synapses*: each synapsis has an associated numerical value, called *synaptic weight*, $w(i,j)$, which represents the *stimulation intensity* of neurone j by neurone i. Conversely, each node (neurone) is characterised by two numerical attributes, its *state* and its *excitation level* (sometimes called the "firing threshold" of the neurone): the state $s(i,t)$ represents a sort of "internal energy" of the neurone, and its effective meaning depends on the problem under consideration; the firing threshold $e(i)$ is the level of $s(i,t)$ which, if reached by neurone i, causes it to "fire" a message to the neurones of the next layer. This "message" represents an energy "quantum" $q(i)$, but it will be received by another neurone j as an energy quantity equal to $q(i)$ multiplied by the synaptic weight $w(i,j)$.

An ANN is dynamically driven by two governing laws: the activation and the iteration law. The activation law is used to update the states $s(i,t)$, while the iteration law rules the order in which the individual neurones are activated. The process starts with all neurones "blank", meaning that $s(i,t=0) = 0$, except for those of the input layer, which are assigned a set of states which represent the context of the problem. At each time t thereafter, each neurone j possesses a *potential* defined as:

$$p(j,t) = \sum_i \left\{ w(i,j)\left[s(i,t-\partial t) - e(i) \right] \right\}$$

This potential is then assigned to the neurone's new state, according to the activation law:

$$s(j,t) = f\langle j,t,p(j,t)\rangle$$

The most commonly employed forms of f are:

- the *identity* law: $f\langle j,t,p(j,t)\rangle = p(j,t)$
- the *unit step* law: $f\langle j,t,p(j,t)\rangle = 1 \text{ if } p(j,t) > 0$
  $f\langle j,t,p(j,t)\rangle = 0 \text{ if } p(j,t) \leq 0$
- the *sigmoid* law: $f\langle j,t,p(j,t)\rangle = 1 / (1 + e^{-p(j,t)})$

This "propagation" from layer to layer causes, after at least n-1 time steps, the "answer" to the problem posed on the input layer to appear as a modification of the states of the output layer: notice that there is no *a priori* assurance that an answer will indeed be generated, because the initial stimulus may well "die out" in the intermediate layers. It is apparent that the response of a neural network to a stimulus (a problem) is heavily depending on both the activation law and the synaptic weights. While the choice of an activation law can be made using common judgement (the three laws listed above have a clear physical significance, see Fig. 3.9) it is impossible, except for very simple or trivial cases, to pre-assign the correct synaptic weights. Here, use is made of a very powerful technique, typical of Neural Networks, which shows extremely promising application capabilities: the technique of self-adjustment of the weights by successive trial-and-error[8]. This self-adjustment technique is an iterative procedure which is started by assigning *two* values to the network: an *initial state* (called "a description of the network universe", or "a stimulus", or "a problem"), and a *target*, which is the answer which we would expect from the network. The response of the network is compared to the target, and, if the discrepancy $\varepsilon =$ (response) - (target) is larger than a prefixed value, *the synaptic weights are automatically modified, according to a pre-assigned protocol*, and the procedure repeated to convergence (the adjustment is made using the derivative of the discrepancy: the implementation details can be found in [15]). This technique is called "self-learning" (sometimes even "self-teaching"!), but it can be seen as being nothing more than a calibration technique, or a non-linear programming, heuristic problem-solving method.

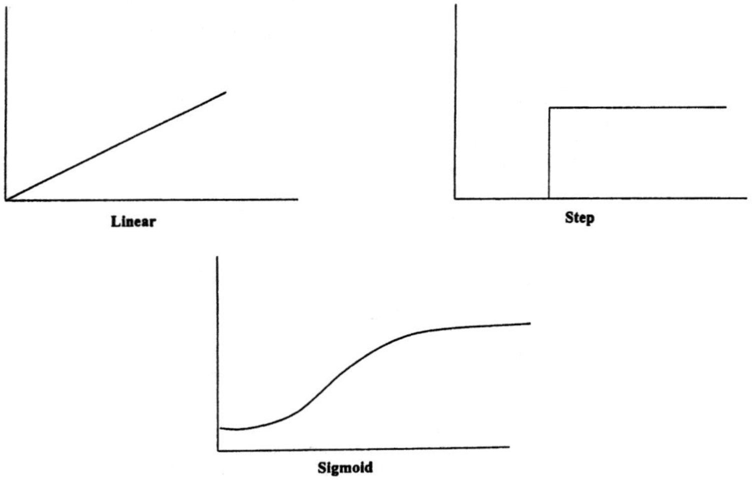

Figure 3-9. Neurone output function characteristics

ANN possess unique characteristics, which make them very powerful tools for Knowledge-Based problems:

a) they can *learn from experience*, not by storing in memory by brute force a large number of instances of resolved problems, but by flexibility, adapting their own internal response to new situations arising in old (i.e., "known") contexts

[8] Notice that this technique is the exact analogue of the procedure supposedly enacted by Biological Neural Systems.

b) they are *interdisciplinary* , in that they can successfully adapt their response to different contexts (because they are not symbol-bound)
c) they can perform with a high level of success in contexts in which the knowledge is imparted in an incomplete, fuzzy, or uncertain way
d) they are robust, i.e., show a very little sensitivity to noisy data.

On the other hand, they lack two very important characteristics of symbolic AI systems:

1. inspectability: ANN -because of the way they operate- are opaque to inspection, in the sense that they cannot produce a *logical explanation* of their response to a certain stimulus. Not only is the logic of their behaviour difficult to inspect (one could have them produce a printout of all state variables at all times, but this would not be easy to check, for a system with -say- 200 neurones); it is also (except for very simple networks consisting of very few neurones) substantially unexplainable, because their response is driven by heuristics, and not by formal logic
2. formal codification: the response of a Neural Network to a given stimulus or group of stimuli cannot be explained in linguistical form (i.e., cannot be expressed symbolically), because its heuristics do not recognise "information bits" in the sense logic -or language- do. Therefore, they are ill-suited for tasks in which the required response is an instruction (e.g., "it is necessary to lower throttle pressure") accompanied by an explanation (e.g., "because load is decreasing and automatic partialization is on").

Typical fields of application for Neural Networks are the following:

- image recognition
- signal interpretation
- voice recognition
- medical direct diagnosing
- diagnosis of malfunctions
- decision support systems
- linear and non-linear optimisation

Of these, only the last three find some applications in the field of energy system design: since actual implementations have been scantily documented until very recent few, the general "public" impression is that they are -for what our field is concerned- an immature technology. This is by no means not true, and we will elaborate a little here, by giving a brief description of some existing implementations. Excluding intelligence- and military applications, where several successful ANN techniques have been applied to pattern and voice recognition, we should consider that any process that can be re-formulated in terms similar to a pattern recognition problem is amenable to an ANN solution. This is especially true for all cases in which the rules and facts or equations which describe the process or the system are either not known or difficult to express in a mathematical or qualitative form: conceptual design, process modelling and process control are typical examples of such problems, where the available knowledge is approximate and/or incomplete, where the amount of data to keep under control is large and where very often there is a lack of models to relate the process parameters to the process performance. We shall separately consider two general areas: Plant- or Process Fault Diagnosis and Process Monitoring.

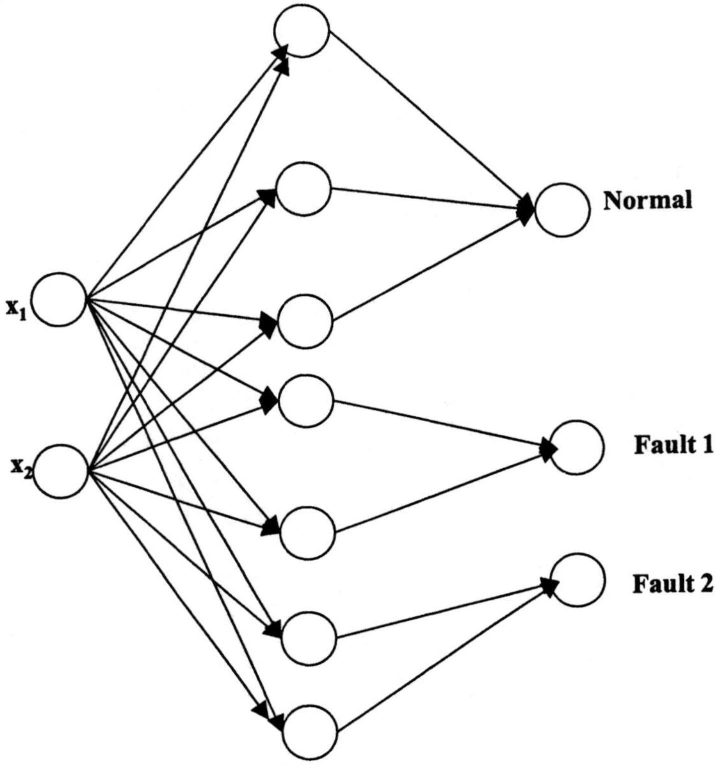

Figure 3-10. Neural Network process representation for fault diagnosis

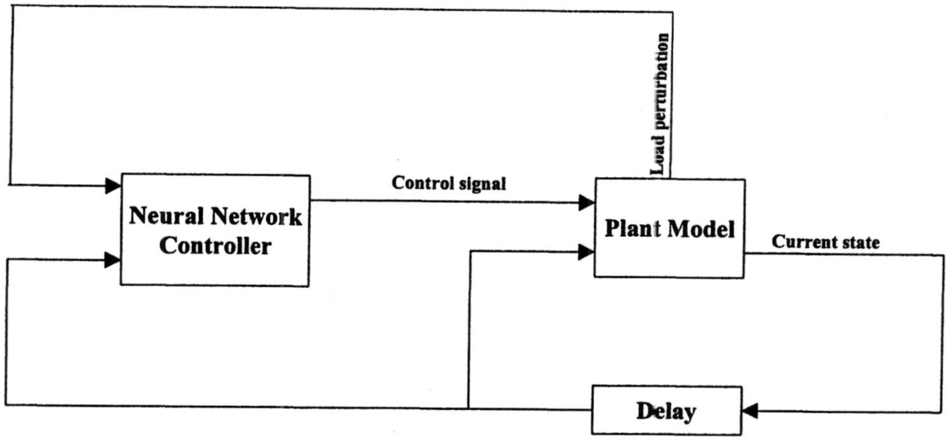

Figure 3-11. Block diagram for a Neural Network controller

Plant trouble-shooting and maintenance are very difficult tasks because a large number of parameters are relevant for the problem solution, and often the data from the monitoring instruments are insufficient and/or not reliable. Operating conditions can gradually change due to degradation of performance, external or internal disturbances, and environmental factors. Detection of malfunctioning is critical for economic and safe operation. In the ANN representation, the input layer can take the input values and the output layer gives the possible normal/fault conditions (Fig. 3.10). A set of simulated measurements are used to *train* the ANN; each set has as a (known) result either a normal condition or a faulty one. The network weights are tuned to reproduce the right results within an acceptable error and with a pre-set value of reliability. If learning convergence fails it is normally necessary to increase the number of the hidden nodes. After the training session(s), the ANN is ready for examining new cases. The correctness of an ANN evaluation can be measured by comparing the model results with the expert answers to the same plant operating conditions; accuracy usually range from 90% to 100%.

Another application of ANN's is the stability control of power systems: when mechanical power is produced by a steam turbine and delivered to a synchronous generator, the frequency and voltage of the current are affected by changes in user power demand (perturbations). For example, if the load suddenly increases the generator shift slows down, and the frequency decreases: the control system must detect the load variation and increase the steam flow rate to the turbine, to increase the turbine mechanical power output. Conventional control systems are somewhat slow and inefficient in handling system non-linearities. ANN-based controllers have delivered better performances. A scheme of such an ANN controller is shown in Fig. 3.11. The load perturbations are the input values and the control signal is the output value: Fig. 3.12 displays the response curves for a conventional- and a Neural Network controller. The ANN-based version has a better response because it makes use of the estimate of the load perturbation as input and is in effect able to immediately counteract its effect, whereas a conventional controller has to wait until the error becomes significantly large before activating the control system response.

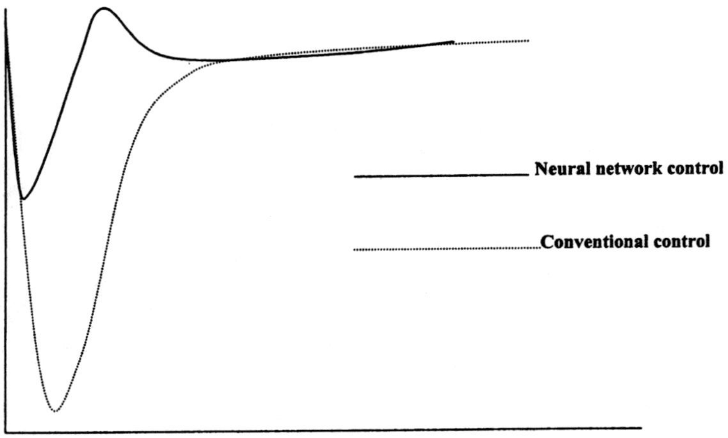

Figure 3-12. Qualitative response for conventional and Neural Network controllers

As a last example, consider a modern chemical process: these processes, and the plants which enact them, are extremely complex and prone both to equipment malfunctioning and operator errors. It is therefore very important to diagnose and eliminate potential errors *before* they may occur. For example, for a system of continuously stirred tank reactors, possible deviations from the correct process are:

- non-exact (meaning "off-spec") inlet concentration values
- non-exact inlet mass flow rates
- non-exact inlet flow temperatures

It is relatively easy to implement an ANN to detect and correlate the errors to the values detected by the measurement chain: the training session consists in this case in assigning first the *desired* target values together with the *correct* inputs, and then the same outputs together with the *incorrect* inputs. The ANN will produce a series of control- and regulating signals that can then be fed to the control system.

Yet another application of ANN has been implemented in the monitoring of a nuclear plant. In this case autoassociative ANN have been used: in these Systems, some of the inputs are represented by the same variables as the corresponding outputs. Back-propagation using a sigmoidal function with an adjustable coefficient was used to train the network, and the output values represent estimates of the final desired values of the operating parameters. Particularly important in this case is the detection of abnormal events. The ANN recognises the "operational state" of the system and detects changes therein, so that it can also predict the output conditions for any combination of variations in the operating parameters.

Finally, the performance of a plant under dynamic operating conditions can be evaluated with ANN: once the Neural Network has been trained it is possible to determine the sensitivity of the output to each parameter. By the same process, it is then possible to optimise the efficiency of the plant.

3.7 - CASUAL VS. MECHANICAL LEARNING. "MEMORY"

"Learning" denotes the ability of a system to modify either its Knowledge Base or its Inference Engine, or both, in response to changes in its environment, so that it can improve its performance with regard to a specific task or a set of related facts (the code "does it better" if it is faced again with these tasks). The most basic way in which a code can "learn" is by being directly instructed by the user via some suitable interface: this has been called "rote learning" [74], and is obviously not very intelligent and very labour intensive. A little more complex is the "learning by deduction", in which the user supplies to the code a list of facts ("causes") and a set of deduction rules, and the code then deduces from these causes the proper effects, under the given rules. This type of learning is useful because it proceeds from the general to the specific, and can be used for instance to "teach" numerical computations to a code (the causes being the data and constraints, and the rules being the operations to be performed on the data).

Some ES (notably, Neural Networks) have the capability of "learning" in a *casual* fashion: in spite of the presence of some degree of meta-knowledge hard-coded into the system, the major part of learning process remains entirely heuristic.

There are other forms of "learning", which by contrast are called *mechanical*: they consist of the repeated application of the principle "if a solution has been found for a certain sub-problem, store it and use it *as is* when that particular sub-problem arises again". This technique is based on the empirical observation that *in scanning decision trees of "similar" problems it is highly likely that "similar" solution paths will be found.* This is the more true, the more "local" the solution is, i.e., the logically simpler -or more elementary- a sub-problem is. This form of learning is particularly well suited to thermal process optimisation problems, in which it is very likely that *local* sub-optimisation strategies can be carried out with identical procedures for whatever plant configuration. As an example, the steam extraction optimisation *procedure* is the same whether the cycle has 2 or 5 or *n* feedwater heaters.

There are yet other forms of learning, the most important ones being the so-called *qualitative parameter-adjustment technique* (or *learning by analogy*) and *the conceptual learning technique* (or *learning by discovery*). In the former, numerical weights are attributed to a certain number *M* of *N* pre-determined user-specified operators which have been selected as "analogy criteria": then, "qualitative differences" are evaluated between initial state and goal, and the weighing formula is varied heuristically to reduce these differences in some norm. A qualitative difference is the equivalent of a partial derivative in "attribute space": a turbine and a pump differ for "type" and "function", while a centrifugal and a volumetric pump differ only in "type", so the δ*(turbine - pump)* is greater than the δ*(centrifugal_pump - volumetric_pump)*. This technique [22,56,70] can become very cumbersome if the number *M* is high: on the other hand, if $M << N$, the results could be very marginal (i.e., the process learns very little). In conceptual learning, which is typical of Object-Oriented Knowledge Systems, certain operators (of the type "IF p AND q AND ...r THEN action") scan the Knowledge Base and infer from it all consequences they can; then, these consequences are added to the Knowledge Base, and the process is repeated every *n* tree-searching moves. This technique is in principle very powerful, but for large problems it becomes unwieldy.

Another technique of conceptual learning is based on a method called *conceptual clustering* [74]: the logical relationship between a solution and the Knowledge Base from which it originated is subjected to a meta-level analysis which tries to *extract logical (meta)inferences from it*; if a meta-rule is found, it is added to the KB, and constitutes a fact which -being not present at all in the previous KB- has been *apprehended* by the code.

Though the entire field of "automatic learning" is being developed at a very fast pace, for what applications to our purposes are concerned (i.e., for design problems), it can be considered still in its infancy. In practice, most "learning" or "self teaching" codes rely on Neural Networks techniques. It is conceivable that in the near future (two- to five years) AI-based design codes embodying a combination of mechanical and casual learning will become commercially available. Applications of conceptual learning and conceptual clustering are still quite far down the road.

"Memory" is strictly connected to "learning", though different enough from it to be treated separately. By "memory" we intend here the capability of a code to remember concepts and facts that were not originally in its Knowledge Base, but were somehow derived by the Inferential Engine in its deductive or inductive procedure. Again, the topic is the subject of rapid development, and it is also specialistic enough to exceed the scope of this Book: it is useful though to present here at least an introductory discussion of the general problem and of some of the different approaches that have been tried. First of all, let us remark that *storing in* memory and *retrieving from* memory are two quite different actions: while storing can be thought of as a completely mechanistic action, very similar in practice to the building of a data base, retrieving resembles rather a strongly pruned search algorithm. Imagine that an "Expert Process Synthesiser" is at work on some design: suppose it finds a certain number *m* of processes P_j (j = 1,2,...m) that meet the specifications, and stores them in its "working memory", ready for being displayed to the user. Since each P_j will be stored as an object, it can be retrieved either as a whole (by "name", so to say), or on the basis of any of its attributes: this is an example of retrieval based on a *direct search by keywords* (the name of the object, say "P_3", being considered as an attribute). This procedure is very simple, strikingly similar to an elementary biological "memory" activity, and can be successfully implemented in practice [42]: it is though limited by its inability to go beyond what is in effect a "labelling" task. If we want to improve the performance of the retrieval action, it is necessary to introduce an inference procedure that can first draw analogies or draw deductions from the existing data stored in the working memory and then accordingly choose the proper keywords: this is tantamount to introducing an auxiliary Expert System into the ES, that drives the search using its own Inferential Engine. Let us consider as an example an important feature of an Expert Process Assistant: the capability to make proper use of waste fluxes and streams. Suppose our hypothetical "Expert Process Synthesiser" is assembling a process, and has already put together a few components in such a way that the raw materials undergo some transformation towards the generation of the products, and assume that some of these components have two or more outputs, one of which is the one we call the "main" output (the "product" of that component): this flux is the one which shall constitute the input of the next component to be added, while the remaining ones will be somehow "available", but are unused for the main production line (the "secondary fluxes" or "by-products"). It would be desirable (because this is indeed the way a human expert would proceed in the design) to keep track of these streams and fluxes in such a way that they can be considered for use as "potential inputs" at some later stage. This can be achieved in principle by constructing an auxiliary ES that performs the following operations:

1. for each component added to the process, stores all the by-products into the Working Memory, as Objects belonging to a Class "Utilities", each Object being identified by its proper attributes;
2. when considering which new component can be added to the existing partially assembled process structure, the fluxes placed in memory as Utilities are scanned first, and those among them which can be used to feed the new component are "recycled" internally;
3. after the process has been completely assembled (i.e., a structure has been constructed which transforms the raw materials into the specified products), the remaining Utilities are inspected again, to see if any of them can be internally or externally recycled to "improve" the process. The improvement can consist in more units of output for unit of input (higher productivity), in

reduced use of external utilities (higher efficiency, lower fuel consumption), or in the contemporary production, in addition to the specified output, of some other commercially or technically desirable "by-product";

4. if several process structures have been found, all of them satisfying the design requirements, the auxiliary ES can even rank them in descending "desirability" order, by considering more desirable those which have a higher productivity, or a higher efficiency, or generate the more convenient by-products.

This example shows that a desirable attribute of a "memory" is its ability of *detecting qualitative analogies* between the objects stored in the physical memory area and those we are searching for. The procedures which perform this task are (quite complex) ES themselves, and the logical operation they perform is called *matching*. Let us denote as "data" the objects stored in memory, and as "targets" those we are looking for. A "match" is said to exist between one piece of data and a target if:

1. the list of the attributes of the two objects is the same. This is the most obvious match, and is called *identity*;
2. all of the attributes of the target are contained in the list of attributes of the data. In this case, a further scanning action is necessary to verify that the target is indeed, in a sense, an instance of the class "data". The attributes of the target are said to be *embedded* in those of the data;
3. there is no immediate correspondence between the two lists of attributes, but it is possible to derive functional similarities, in the sense that there exist meta-levels (at least one!) at which the two lists share some meta-attributes.

In cases 1 and 2, the target is similar to (in case 1, identical with) the data, and retrieval is immediate. In case 3, a further intelligent search is required at the detected meta-level. Notice that the keyword search can be included in case 3 (and the meta-level is identified by the chosen keyword).

In the field of AI at large, extensive research on "memory" applications is being performed in the fields of image- and sound recognition: while the developments are -at present- in no way directly applicable to the field of Process Design, it would be advisable that interested readers consult the monographs on this topic [2,3,19,27,60,80].

3.8 - Approximate Reasoning. Belief

To "apply" a rule to a portion of the Knowledge Base involves two different tasks:

1. deciding whether the state described by the knowledge batch (the *facts*) of interest is comprised in, or in the limit coincides with, the domain of that rule (see Section 3.1.6);
2. applying the rule and adding its co-domain to the Knowledge Base.

The most obvious way of executing step 1) is to scan *all* rules and compare their domain with the facts under consideration: but, for large databases and complicated inference engines, this can become a very taxing task. The problem is most efficiently solved via a hierarchical search which *starts from the given facts* and eliminates successively all rules which do not contain those facts *and all rules which can be derived from these*. The process must be organised hierarchically, so that it first eliminates single rules and then rules derived from them, because there can be rules which *could* be derived from a rule which does not apply to the given facts, but which *could also be directly derived from the facts*. Notice that the validity of a procedure like the one just described is based on the necessary assumption that both the given facts and the applicable rules are well-formulated and exactly known: if either the facts or the domain of a rule are *vague* or *approximated*, it is obvious that no rule can be eliminated without first resolving the approximation or the vagueness of its domain or of the facts which lead to its elimination. This means that additional complexity and flexibility is required of the solution procedure: typically, we would need an additional set of rules which tells us how to match "approximate" facts. These rules expand the approximate descriptions to specify all of its implications. For example: "steam near its saturation line" can be made explicit by rephrasing it as "water in a state such that (T) GREATER THAN (T_{sat}) AND (T - T_{sat}) LESS THAN ε, with, say, $\varepsilon \leq 0.5$ °K. This is, in essence, a *syntactical analysis of the given facts*, and, though in principle always successful, it is very cumbersome to implement.

A particular meta-rule which makes coping with approximate knowledge representations easier is the "*keyword rule*" (see also above, Section 3.7). It is again a hierarchical rule, which considers *only certain "key" features of the given facts to be important*, and tries to match them with the domains of the available inference rules. This is a very powerful approach, provided the choice of the "keywords" is proper, which is of course a problem in itself (which ought to be solved by the Domain Expert).

Besides syntactical and keyword analysis, there are procedures derived from mathematical theories that allow to compute the conditional probability of a correlated set of events: these are the so-called Bayesian Conditional Probability Method, the Certainty approach, and the Dempster-Shafer theory. The description of these applications and of the underlying theories is outside of the scope of this book, and interested readers are referred to the excellent treatment presented in [74]. What we can say here though is that all of these methods suffer from a fundamental weakness: they rely, in one way or another, on *quantitative measures of the degree of uncertainty which can be attached to a certain proposition*. These measures are usually expressed by real numbers between -1 and 1 (-1 representing absolute unlikeness and 1 absolute certainty), and can be assigned only by Domain Experts: nevertheless, it has been found that they tend to cover a sort of foggy area in the interval [-1,1] rather than assume a precise numerical value (e.g., 0.4). Furthermore, the quantitative measure of this uncertainty is strongly dependent on both the immediate and remote experience of the Domain Expert, and on the physical and logical environment of the case under study. In the absence of better methods, we find that Fuzzy Set Theory (Section 3.5 above) is still an engineer's best choice when dealing with approximate knowledge.

A different type of probem is posed by *ambiguous knowledge*. Ambiguity differs from vagueness: probably the best definition is given by [55]: "ambiguous terms may be at once clearly true of some objects and clearly false of them". It can be shown that ambiguities are in fact invariably linguistic (semantic) problems, and therefore they should be resolved in the phase of "Problem Formulation".

Another common type of knowledge, closely related to approximate reasoning, is that expressed under the form of *belief*. It is approximate, because it does not result in "IF p THEN q" statements, but it turns out it can be treated in a very effective way. Without going into details, probably the best way of treating knowledge expressed under the form of a *believed fact* (e.g.: "I believe that the number of feedwater heaters in a conventional steam plant never exceeds eight") is that of translating into a form of *conventional knowledge about the believed fact and some unlikely or unusual counterfact* [73] (e.g., "EITHER (number of feedwater heaters) ≤ LESS-OR-EQUAL-TO 8 OR probability(plant is a conventional steam plant) ≤ 0.01"). It is important to remark here that not all such translations have to recur to probabilistic utterances: what is of essence is that the fact which would disprove the belief be identified by a quantifiable degree of unlikeliness.

The reason why we should seek ways to handle approximate knowledge is that *it is much more powerful than exact knowledge*: it can allow for inference in cases in which an exact treatment would fail. For example, when designing a Brayton-based gas turbine plant, we may know in advance the maximum allowable cycle temperature, $T_3 = T_{max}$ in Fig. 2.4, but need to know the pressure ratio and the inlet temperature T_1 to compute the required fuel-to-air mass ratio, α. If the facts we assign are "T_1, p_1, p_2, α", the code would be unable to cope with situations in which T_3 is different from the specified T_{max}. But if an *approximate* fact is substituted for , like "$0.01 \leq \alpha \leq 0.05$", the domain of application of the code is substantially increased. Similarly, approximate facts are very useful in monitoring and control applications.

3.9 - SEARCH METHODS

As repeatedly stated in the previous sections of this Chapter, knowledge-based methods are in essence search methods which try to scan a (possibly very large) solution space to find the or at least a solution for the problem under consideration. Search techniques therefore, and in general the problem of the most appropriate search method, are a central issue in AI. A "search problem" is characterised by a description of an initial and of a final state, not necessarily represented at the same abstraction level. A "state" is here to be intended as the description of a well-identified situation, or more precisely as the description of a *structured objective* situation. The search is conducted by means of logical operators which, when applied to a certain state (a situation) transform it into a different state. These operators can be logical inference rules, representations of physical laws, or purely conventional *ad hoc* rules considered valid for that particular search. Two sets of states are associated to each of these operators: the *domain space*, defined as the set of all states to which the operator may be applied, and the *co-domain* (or *projection*) *space*, which is the set of all states which may be generated by the operator. If an operator **O**, applied to a state **P**, transforms it into a state **R**, then **R** is said to be **P**'s *first successor*, and correspondingly **P** is said to be **R**'s *first ancestor*. There can be second, third, ... ancestors and successors: given an initial state I, the set S composed of all of its successors is called the *search space*, and obviously corresponds to the solution space, i.e., *if a solution exists, it is a member of this set.*

There are several "search strategies" or "techniques", among which:

- uniform irrevocable search strategy: "Hill climbing" (all the so-called "gradient techniques" belong to this class);
- uniform, "depth first" search, "LIFO" ("Last-In-First-Out": the last state which was identified will be the first one to be operated on);
- uniform, "breadth first", "FIFO" ("FIrst-In-First-Out": the first state which is identified will be the first one to operate on);
- heuristic (possibly, weighed) search for the first local optimum.

These algorithms are well known, and several practical implementations have been published: interested readers are referred to [49], which contains detailed descriptions and macro-code schemes for many search procedures.

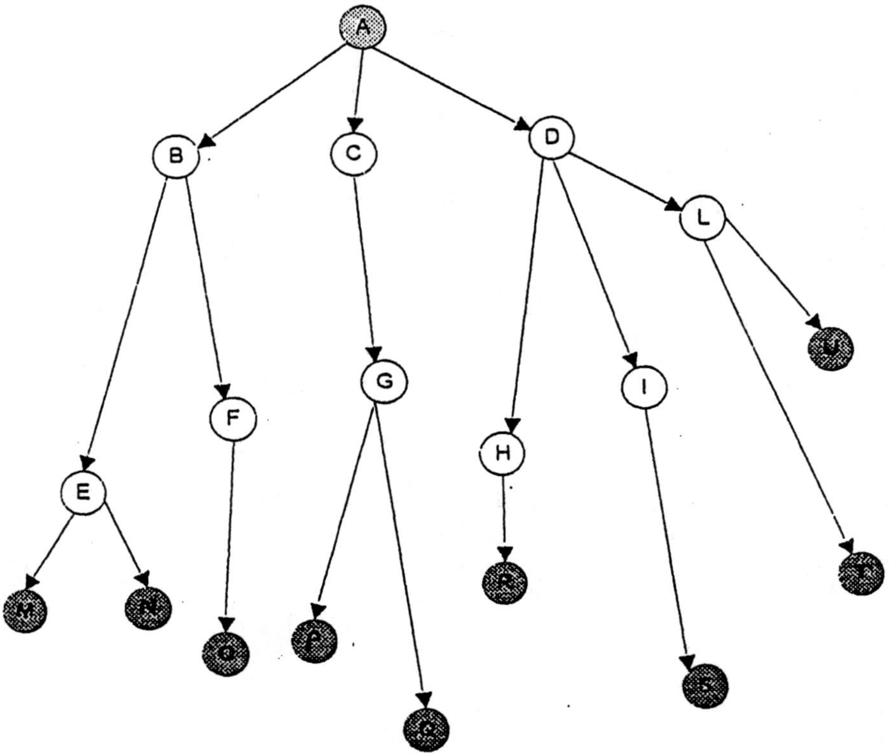

Figure 3-13. Decision tree for a multiple-choice situation

3.9.1 AUTOMATING SCANNING OF A DECISION TREE

The process of searching for "complete paths" within a decision tree can become exceedingly taxing for large trees. Therefore, it is useful to have access to *tree-scanning strategies* which facilitate this task. We will explore the two most popular methods for searching for the termination of each branch: the *search in depth* and the *search in breadth.*

Consider the tree depicted in Fig. 3.13, and make the following positions:

- **A** is the search space, i.e., the entire tree consisting of **N** nodes and **E** branches
- node **A** represents the "initial" state
- there is a certain well-defined subset **D** of the other nodes consisting of all the nodes which represent "known" states of the system (either by measurement,

by experience or by specification): in the present example, this subset comprises nodes N,O,Q,R, and T

"To search" the tree means then to start from node **A**, and try to determine *all* of the branches connecting **A** to one of the nodes belonging to **D**. As a practical application of a search procedure to the tree depicted in Fig. 3.13, consider for example the case in which a process has "failed" (i.e., malfunctioned) under a certain set of operative conditions identified by the point "**A**" in its operational space, and that all nodes of the tree represent some failure condition (e.g., "lubrication oil temperature too high", or "water level too low", etc). The nodes N,O,Q,R,T of the "known" subset **D** represent here the last measurements taken before failure, and work in a sense as constraints in the diagnostic procedure.

A *search in breadth* will scan all of the nodes j ($j = 1,2,3...M$) at a certain level i before the next level is processed: it can be performed by a procedure whose macro-code looks like the following:

```
order A according to any numbering, call the ordered set "OA"
FOR (all nodes in OA)
 WHILE (OA not empty) DO
    SET N_i = First node in OA
    IF (N_i belongs to D) THEN
       IF (diagnosis for N_i accepted) THEN EXIT
       END IF
    END IF
    SET OA = OA - N_i
    FOR (successor S_i,j+1 of N_i) DO
       OA = S_i,j+1 + OA
    END FOR
 END WHILE
END FOR
```

The order of the scanning generated by the above procedure is shown by the numbers attached to each branch of the tree in Fig. 3.14.

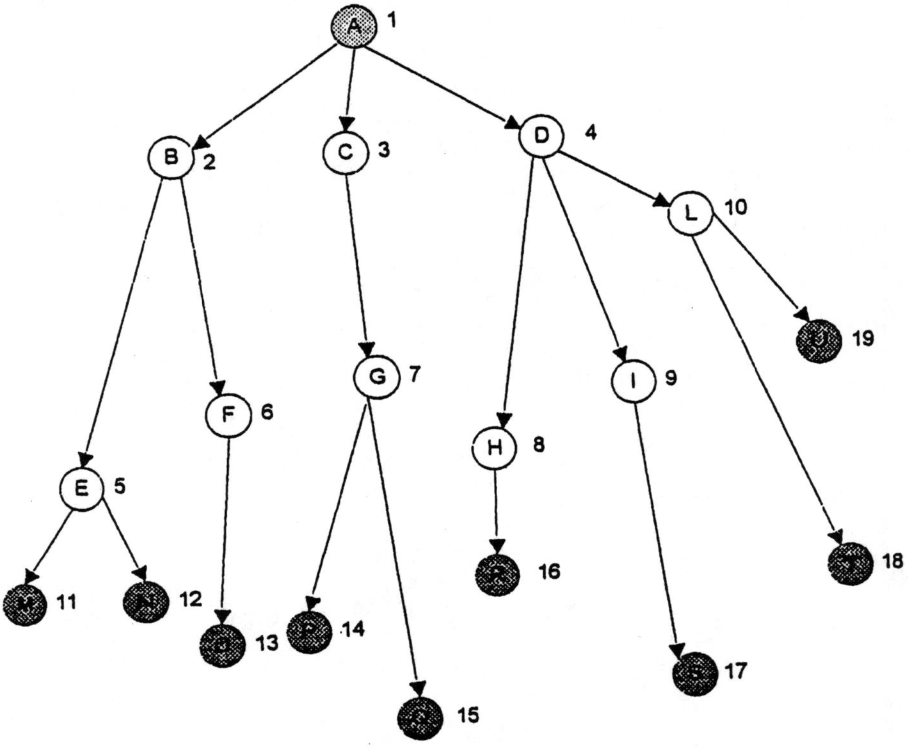

Figure 3-14. Breadth-first search

A *search in depth*, on the other hand, starting from the same node N_1, will scan all of its successors $S_{i+1,j}$ before proceeding to the adjacent node $S_{i,j+1}$ at the same level. It can be performed by a procedure with the following macro-code:

```
order A according to any numbering, call the ordered set "OA"
FOR (all nodes in OA)
  WHILE (OA not empty) DO
    SET N_i = First node in OA
    IF (N_i belongs to D) THEN
      IF (diagnosis for N_i accepted) THEN EXIT
      END IF
    END IF
    SET OA = OA - N_i
    FOR (successor S_{i+1,j} of N_i) DO
      OA = S_{i+1,j} + OA
```

12. END FOR
13. END WHILE
14. END FOR

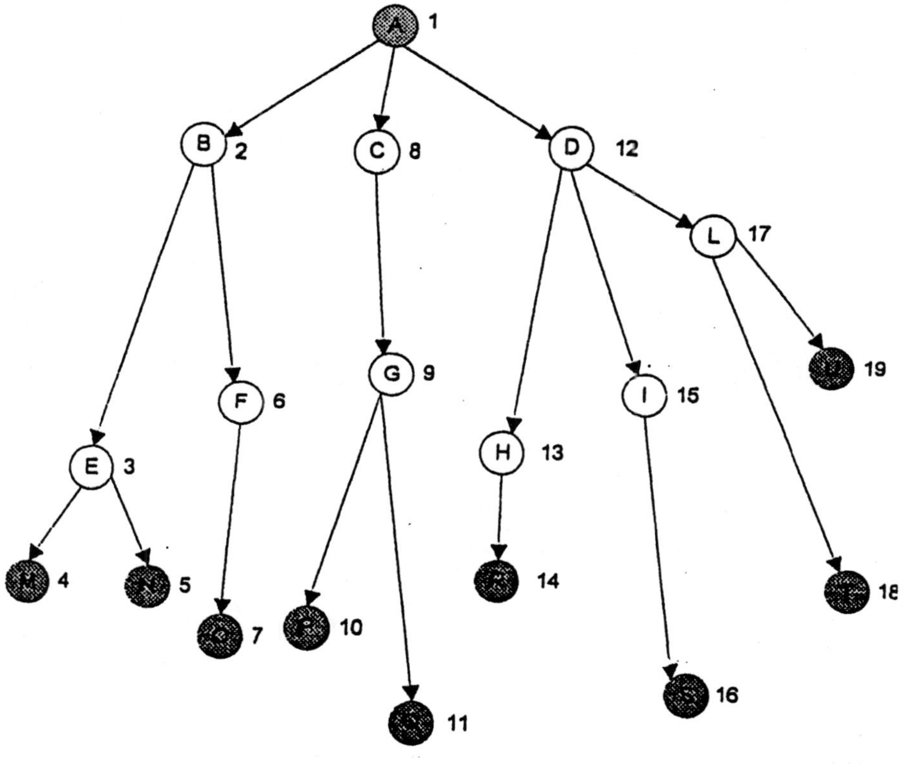

Figure 3-15. Depth-first search

The order of the scanning generated by this procedure is shown by the numbers attached to each branch of the tree in Fig. 3.15. To expand on the "tree" analogy, it is clear that the search in breadth scans all branches first, then all twigs, then the smaller twigs, then the leaves. A search in depth selects a branch, scans all of its twigs one by one by examining all the leaves for each twig, and then proceeds to the next branch. It appears that the search-in-depth gives *some answers* after very few scanning steps (i.e., it reaches nodes M and N after respectively 4 and 5 steps), while the search-in-breadth gives all the answers sequentially at the end of the scan: this difference may become of importance when the scanned tree possesses a high number of nodes.

3.9.2 WEIGHTED SEARCH OF A DECISION TREE

Consider the tree of Fig. 3.16, which represents the engineering production chain of a certain product (in this case, steam). In such graphs, the nodes are the "states" of the product in the course of its processing, and each branch represents a process. Assume that the numbers assigned to each branch represent *the dimensionless value of an engineering relevant extensive parameter* (a mass flow, an exergetic cost, a monetary cost, a resource consumption, etc.) *that can be used as a measure of the "merit" of that process*: we are interested in devising a procedure which searches the tree to find the path from node 0 to node 1 (from the final product to the raw materials in the environment) *which has the minimum [vz maximum] value of the performance parameter.*

The search algorithms are the same as those discussed in the previous example: this time though, before leaving each new node reached through the path, the value of the "weight" (as the performance parameters are often called) must be updated. To this aim, the following lines can be inserted between steps 11 and 12 of the procedures outlined in Section 3.9.1:

1. FOR ALL (AND branches) DO
2. WEIGHT = WEIGHT + WEIGHT ($AND_{(i,i+1)}$)
3. END DO
4. FOR ALL (OR branches) DO
5. WEIGHT = WEIGHT + WEIGHT (AND_j)
6. END DO

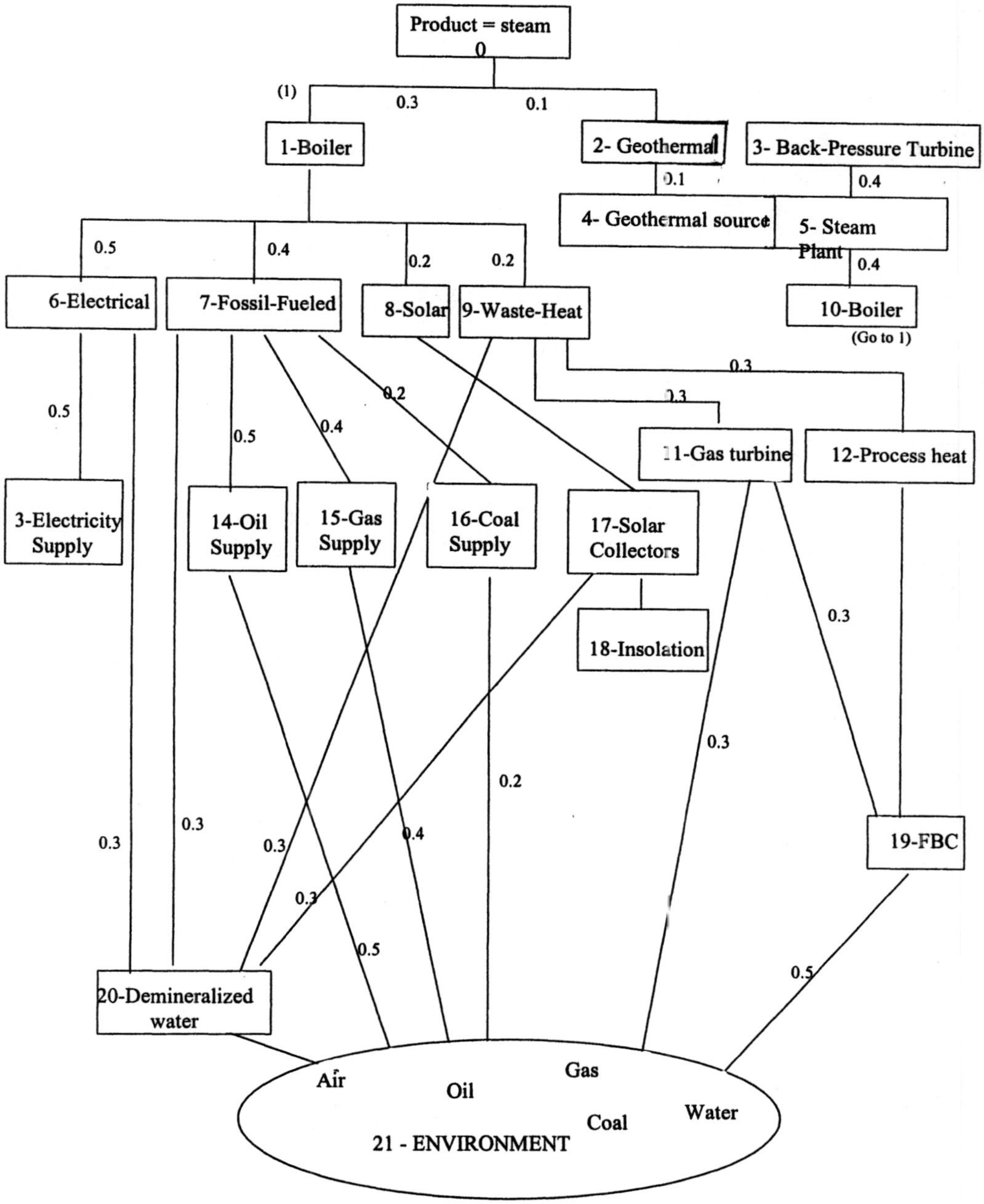

Figure 3-16. Tree representation of a steam generation line

In practice, searches of this kind can become prohibitively expensive in terms of computer resources if the tree has a very high number of nodes (like, for example, in many chemical or mechanical industrial processes). Restricted search strategies are available to handle these cases [3,53,80]. A different approach consists of using an Object-Oriented approach to the search: suppose we can be that each node and branch as an Object, and translate all of their attributes in this new syntax. Now, let us introduce an additional attribute, a *preference index* π, which can be defined here as the reciprocal of the weight: the higher the weight (penalty), the lower π. So, the original tree is now replaced by a series of Objects (each labelled by its own π), linked by some relationships expressed in the form of rules, so that a path from the root to a leaf can be reconstructed on the basis of a series of necessary inferences of the type "IF N_j THEN N_k." A weighted search can be conducted in our Object space by a procedure like the following:

1. FOR (all nodes in **OA**)
2. WHILE (OA not empty) DO
3. SET N_i = First node in **OA**
4. IF (N_i belongs to **D**) THEN
   CONSTRUCT INFERENCE CHAIN C_{OA} (from N_i to N_M)
5. COMPUTE $\Pi = \sum_{i=1}$.
6. IF (Π = max of all previous chains) THEN $C_{OA} = C_{opt}$
7. END FOR
8. END WHILE
9. END FOR

The above procedure is based on the concept of "logic with priority", which is a non-monotonic logic in which the decision is made not only on the basis of rules, but on a joint basis of (rules + priority parameter) [20]. Object-oriented procedures are convenient for weighted searches of either extremely large or unknown trees, where it can be easier to produce relationally linked lists than root-branch-twig-leave paths.

3.9.3 Reduction of a Tree to a Purely AND/OR Tree

Consider the decision tree depicted in Fig. 3.17: it possesses some "AND" nodes (D,E), some "OR" nodes (B,F,G,L) and some hybrid nodes (See Section 3.1.16). A possible algorithm for *reducing* a hybrid tree to an AND/OR tree is sketched here below.

1. FOR (all nodes N_i in **OA**) DO
2. FOR (all successors) DO
3. IF (S_j is an AND node) THEN

```
        CALL (insert an AND node NS_AND between N_i and S_j)
    END IF
    OA = OA - N_i
    END FOR
  OA = NS_AND + OA
END FOR
```

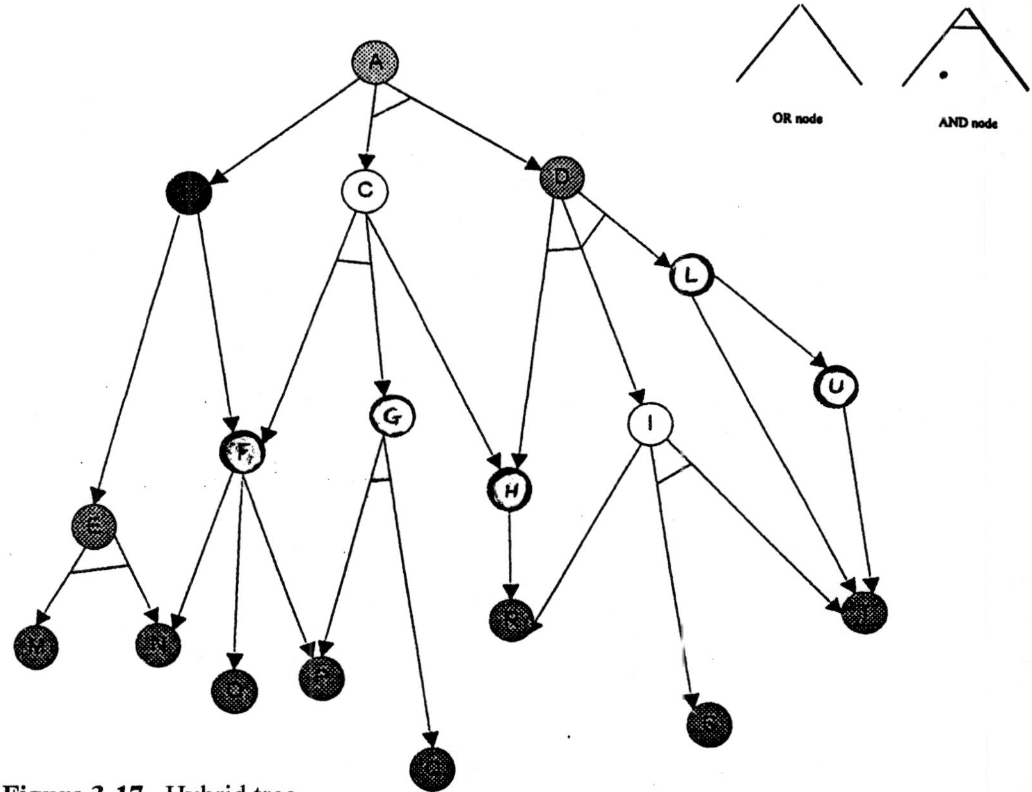

Figure 3-17. Hybrid tree

The concept is simple: each time a hybrid node is encountered, it is reduced to an OR node by inserting a virtual AND-node NS between it and its AND branch. The virtual nodes do not correspond to any physical state, but are only convenient for calculation purposes. The transformed tree is shown in Fig. 3.18.

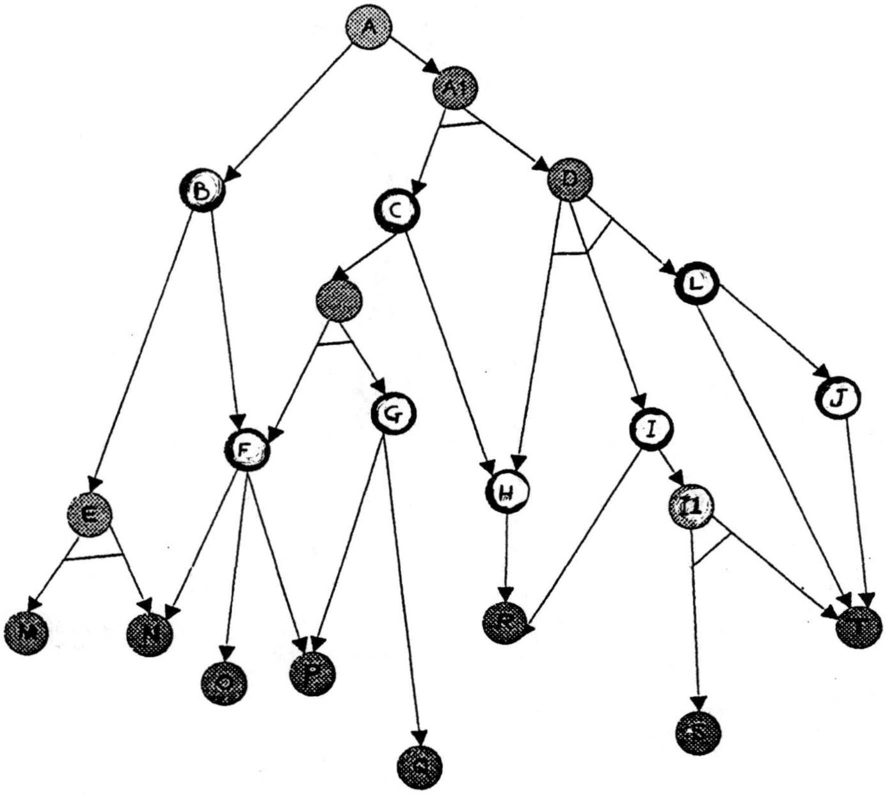

Figure 3-18. The reduced tree (notice the AND/OR structure)

3.10 - HANDLING OF CONSTRAINTS

Every engineering design task is subject to a series of physical, technological, procedural, environmental or legal requirements that limit the choice of system configurations: often, these restrictions are so severe that they direct the designer towards a specific set of configurations; and at times they are so strict that this set reduces to one configuration only. When the Design task is formulated as an Engineering problem, these "restrictions" become *constraints*, which must be accounted for by the design paradigm. Constraints can be subdivided in four fundamental categories:

- *causal constraints*, which express conditions of equilibrium, mass- or energy balances, etc. These are known *a priori*, and can be embodied directly into the design procedure;
- *parametric constraints*, which represent size or technological limitations, etc., and are also known *a priori* or established in the initial phases of the design activity. They, too can be embodied directly into the design procedure;
- *conceptual constraints*, which represent the global and specific experience of the design team, and guide them in their design choices. These constraints ought to be included in the Knowledge Base;
- *interaction constraints*, which arise from the interactions among different subsystems, and are due to compatibility between components or between a component and certain fluids, to the requirement of matching thermodynamic properties between the output of one component and the input of the next one, etc.. These constraints are difficult to incorporate as rules or facts in the Knowledge Base, because of the multitude of the possible occurrences.

Regardless of its category, a constraints can be numerical or symbolic. The meaning is obvious, and it is also obvious that a symbolic constraint is under all respects a *rule*, and can be treated as such by the inference engine. Numerical constraints can be divided in *strong* ("value of attribute i of object p" = "value of attribute j of object q") and *weak* ("value of attribute i of object p" $\geq$ "value of attribute j of object q"). Usually, it is convenient to impose weak constraints in their strong formulation (replacing the inequality by the equality sign), thus delimiting the general area in the (i,j) plane where the solution must be sought; at a later stage, a refined search can be conducted to consider the effects of the weak relaxation of that specific constraint.

It is important to consider that constraints can be usefully coupled with searching-algorithms and help "streamline" the tree-scanning activity: if symbolic constraints are "added" to the relevant nodes (or branches) as additional evaluation rules, the search paradigm can for example skip all of the nodes which are not "affected" by a certain rule, thus resulting in a more effective pruning of the tree. If the constraint is numerical, its arguments can be normalised (i.e., reduced to a convenient dimensionless number between 0 and 1, or -1 and 1) and used as nodal or branch weights that constitute an "on-line" evaluating criterion for the scanning algorithm (see Section 3.9.2 above).

3.11 - BACKTRACKING

In a very broad sense, all design procedures can be viewed as a search in the solution space of an "optimal" system that satisfies the design constraints and abides to all constraints. So, decision trees and the relative search methods are important tools in every AI procedure aimed towards the solution of engineering tasks. Since complex tasks generate complicated decision trees with a very high number of nodes and branches, it is of paramount importance to avoid performing extensive searches (i.e., searches in which all of the nodes/branches are visited by the scanning algorithm). The techniques described in Section 3.9 take this into account and, irrespective of whether they are performing a search in depth or in breadth, they stop the scanning, i.e., they do not examine the successor of the node they are currently working on if there exist conditions, either on the next node or in one of the branches that lead to it, that are incompatible with the design goals and/or constraints. The logical problem we want to tackle in this section is then the following. Suppose the scanning algorithm finds that the "solution branch" it is presently working on must be terminated at node *j* of level *i*: where ought the scan to be resumed, i.e., to where ought the procedure to *backtrack*? Depending on the type of search being conducted, the "arithmetical" choice for backtracking should be:

- go back to the last node examined, if the search is a hill climb;
- go to the parent of node *j* at level *i-1*, if the search is in depth;
- go to node *j-1* at level *i*, if the search is in breadth;

But these arithmetical choices (also called "chronological", [74]) are not always the best ones, and some times they actually lower the efficiency of the scanning algorithm: consider for instance the situation depicted in Fig. 3.19, in which a depth search is being executed on a tree (for the procedure, see also [11,17] such that some of the events at level *i+1* are incompatible with their grand-parent at level *I-1*.

Chronological Backtracking: search order

1 - 2	(surface heat exchanger has higher priority)
1 - 2 - 6	(shell & tube heat exchanger has higher priority)
1 - 2 - 6 - 10 - 12 - 16	(depth first - fails: not enough heat exchanged)
	(backtrack to 12)
1 - 2 - 6 - 10 - 12 - 17	(fails: not enough heat exchanged)
	(backtrack to 10)
1 - 2 - 6 - 10 - 13 - 18	(fails: not enough heat exchanged)
	(backtrack to 13)
1 - 2 - 6 - 10 - 13 - 19	(fails: fluid 2 corrodes tubes at this temperature)
	(backtrack to 6)
1 - 2 - 6 - 11	(fails: fluid 2 corrodes tubes as well)
	(backtrack to 2)
1 - 2 - 5	(fails: corrosion problem persists)
	(backtrack to 2)
1 - 2 - 4	(fails: corrosion problem persists)
	(backtrack to 1)
1 - 3 - 7	OK

Dependency-Directed Backtracking: search order

1 - 2	(surface heat exchanger has higher priority)
1 - 2 - 6	(shell & tube heat exchanger has higher priority)
1 - 2 - 6 - 10 - 12 - 16	(depth first fails: not enough heat exchanged)
	(backtrack to 12)
1 - 2 - 6 - 10 - 12 - 17	(fails: not enough heat exchanged)
	(backtrack to 10)
1 - 2 - 6 - 10 - 13 - 18	(fails: not enough heat exchanged)
	(backtrack to 13)
1 - 2 - 6 - 10 - 13 - 19	(fails: fluid 2 cannot be in contact with tubes at high temperature)
	(backtrack to 12)
1 - 3 - 7	OK

Figure 3-19. Backtracking procedures

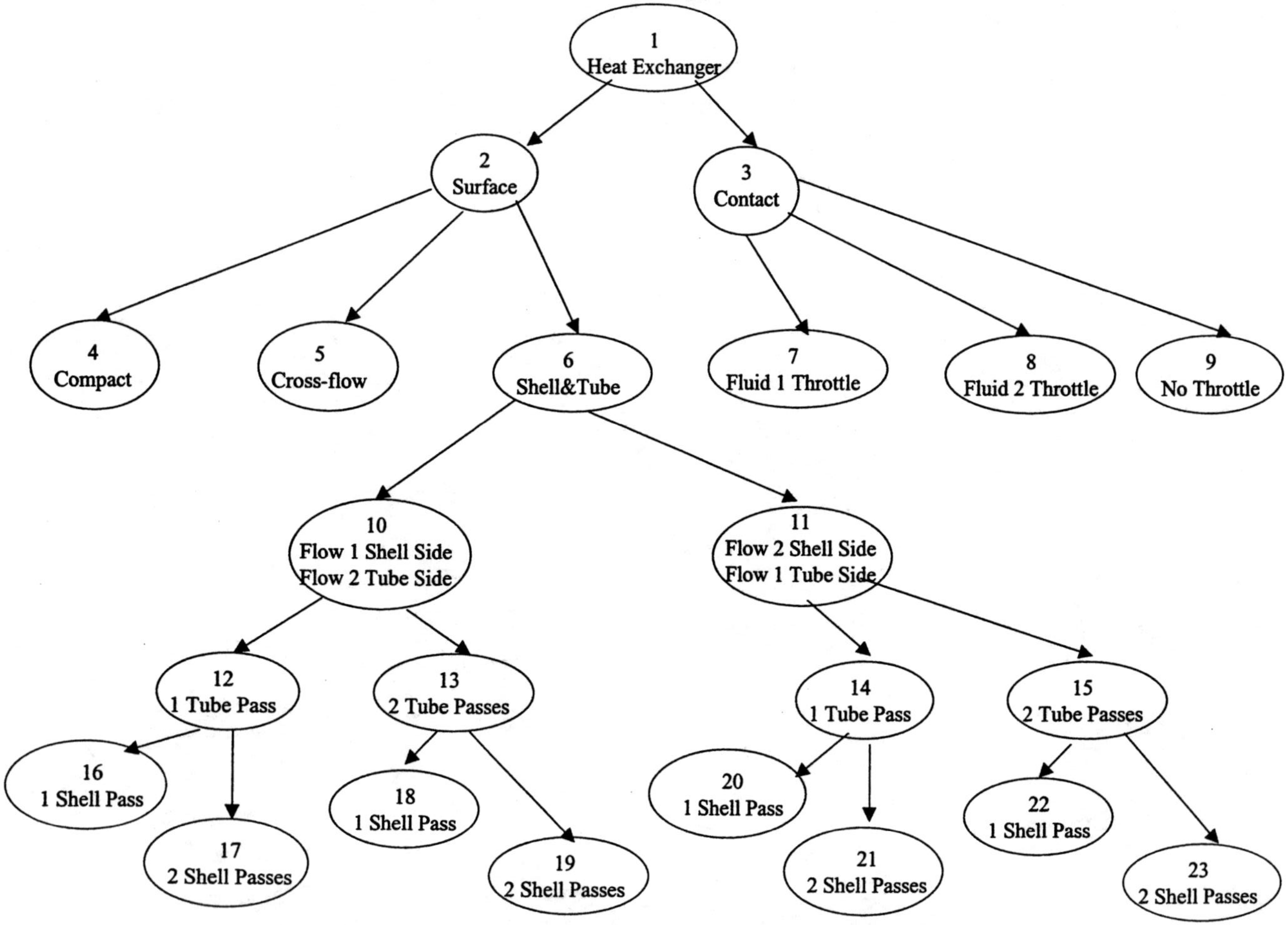

Figure 3-19 (continued). Backtracking procedures

If node $(i+1,1)$ is discarded as unfeasible, and the algorithm backtracks to $(i,1)$, the next node examined will be $(i+1,2)$: if this fails as well, node $(i-1,1)$ will be examined again, and the search will take branch B to node $(i,2)$ and then $(i+1,2)$ and possibly $(i+1,3)$, all of which are incompatible with their grand-parent $(i-1,1)$. Before node $(i+1,4)$ is reached (which represents an acceptable solution), useless (i.e., *a priori* hopeless!) searches have been conducted, and the efficiency of the scanning algorithm is clearly unsatisfactory. Similar situations may be encountered in breadth searches as well. There are even cases in which chronological backtracking incorrectly originates apparent dead-ends. To remove this limitation, a different sort of backtracking, called Dependency-Directed Backtracking ("DDB"), has been proposed (see [74] for reference). In DDB, the rules which led to the successful processing of all antecedents (from primitive ancestors down to grand-grandparents, grandparents and parent nodes) are kept in a list called Dependency Records: if, as in the previous example, a contradiction is detected at node $(i+1,1)$, the search resumes from the *logical antecedent*, i.e., from the nearest ancestor node at which there is no contradiction, $(i-1,1)$ in this case, and furthermore, nodes $(i+1,2)$ and $(i+1,3)$ will not even be considered, because their attribute list contradicts the Dependency Records. This technique is costly in terms of implementation, but its application substantially improves the performance of the scanning algorithm.

3.12 - EXAMPLES AND PROBLEMS

3.12.1 - CONSTRUCTION OF A DECISION TREE

Consider the problem of "*steam generation*". This can be well regarded as a very basic, apparently elementary, design assignment. The (engineering) reasoning needed to solve the problem is well established, because after all steam can be generated by a very limited number of processes: therefore, this example offers a neat possibility of application for a *tree decision diagram*, which is represented in Fig. 3.16 and will be developed here.

One can consider the *object* "STEAM" to be the "fruit" of the tree (as in the scheme of Fig. 3.16), or alternatively its "root" (as it would be if the scheme of the figure was read upside-down). For our purposes this makes no difference, so let us think of "STEAM" as the only fruit of the process tree.

Steam can be found in nature under the form of geothermal eruptions (or, more precisely, it can be generated from a geothermal source): so, this is one natural possibility for steam production, and correspondingly, we insert a box labelled "GEOTHERMAL" *below* the "final product" STEAM.

But steam can also be extracted from a back-pressure turbine, or generated in a boiler: therefore, there are two other boxes in parallel with the "GEOTHERMAL" one, labelled respectively "BOILER" and "BACK-PRESSURE TURBINE".

Upon closer examination, it appears that the geothermal production is only possible if a geothermal source is available: therefore, the *branch* presently ending with "GEOTHERMAL" will have at its *root* by necessity "GEOTHERMAL SOURCE". Moreover, this branch is clearly terminated here, because the logical path "GEOTHERMAL SOURCE⇒GEOTHERMAL CONVERSION⇒STEAM" is a complete (and logically permitted) one. Whether this would be an engineering choice, depends on other factors, not considered here.

The branch which starts with "BOILER" is clearly the most important here: the first choice which is presented to us is that between *electrical*, *fossil-fuelled*, *solar* and *waste-heat boilers*. These four boxes will be four "branches" which collectively bear only one stem ("BOILER") which in turn bears the fruit ("STEAM"). Each of these branches requires some means of energy input, and some flow of properly treated WATER which can be safely evaporated into STEAM. By similar reasoning, the meaning and the position of the remaining boxes is readily determined. It is noteworthy that the *roots* of the decision tree are all materials found in the environment. This does not happen by chance: *every* engineering tree of this kind, which traces back the possible alternatives for the production of some good or service, *is bound to have its roots composed by elements of the domain "environment"*. See Examples 6.6 and 6.7 for some interesting applications of this idea.

The branch which starts with "BACK-PRESSURE TURBINE" must have, as next supporting branch, the "STEAM PLANT" the turbine belongs to; and, since this plant will perforce possess a boiler, a recursive call can be made to the "BOILER" branch discussed above.

The use of such a simple decision tree is trivial: each complete path from the "fruit" STEAM to the "roots" WATER, AIR, FUEL, etc. represents a *logically complete production chain*. The engineering feasibility of each chain can be assessed by assigning some kind of quantitative information ("weight") to each of the boxes ("nodes") and of the branches ("paths"): one speaks of "weighted" or "optimal" search (i.e., tree-scanning) strategies [27,51,53,74].

3.12.2 APPLICATION OF A FUZZY SET TECHNIQUE

Consider the following problem: "devise a computer-based procedure which can give a qualitative, critical estimate of the probability for thermal creep to cause failure of a gas turbine blade. The input decision variable is the gas temperature". Binary logical "thinking" will not suffice here to solve the problem: in Normal Set theory, *either* the blade temperature is such that the blade belongs to the "CREEP-FAILURE" class, *or* the blade belongs to the "NO-CREEP-FAILURE" class. In other words, these classes are mutually exclusive, as of course is binary logic. A form of reasoning which can help us here is the so-called *fuzzy reasoning* (Section 3.5), which can be represented in the following macro-code:

```
READ (gas temperature, T_gas)
IF (blade is cooled) THEN
    (T_blade) = α(T_gas)     [α < 1]
END IF
(fuzzy weight) = f(T_blade)          [f is read from Fig. 3.19]
IF (fuzzy weight ≥ 1) THEN
    FAILURE = CERTAIN
    ELSE IF (fuzzy weight > 0.67 and < 1) THEN
      FAILURE = VERY LIKELY
    ELSE IF (fuzzy weight > 0.33 and < 0.67) THEN
      FAILURE = POSSIBLE
    ELSE IF (fuzzy weight ≤ 0.33) THEN
      FAILURE = VERY UNLIKELY
END IF
```

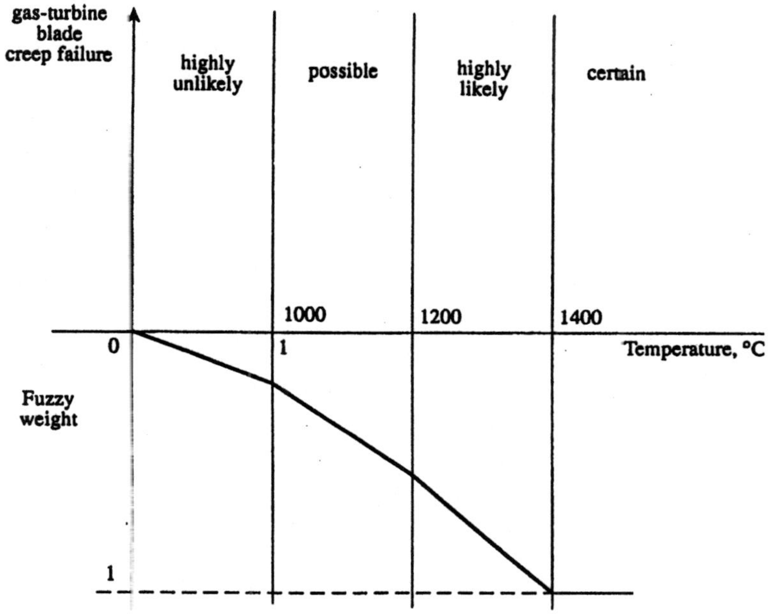

Figure 3-20. Example of the application of a fuzzy weighing function

Albeit this is only a very simple example, all the phases of the fuzzy processing appear here: some deterministic calculations (steps 1-4) fuzzification of the available data (step 5) and (in this example, qualitative) defuzzification with translation of the results of the calculation into non-binary language (steps 6-14).

CHAPTER 4

KNOWLEDGE-BASED METHODS IN THERMAL SYSTEMS ENGINEERING

4.0 - INTRODUCTION

As repeatedly stated in the preceding Sections, scope of any KB method applied to an engineering task is that of assisting -or substituting- the "expert" engineer in performing tasks which can be suitably represented in a form which allows for knowledge manipulation on the part of a computer code, called the Expert System. Before launching the coding activity, the Knowledge Engineer should consider the following questions:

1. How shall the knowledge necessary for expert-level performance be represented as symbolic data structures for computer use?
2. How can one achieve flexibility in adding and changing knowledge in the development of a Knowledge Base?
3. How can the knowledge in a specific field be systematically acquired for computer use?
4. Can it be discovered by a program itself, acting autonomously?
5. What designs are available for the inference procedure to be used by the intelligent artifact?
6. How can one achieve efficient and accurate performance as the combinatorics of the problem space increase?

If the power of an AI program is primarily a function of the quality and completeness of its knowledge base, then these are the critical bottlenecks of KBS. Four recently formalised basic principles [60] ought to guide the Knowledge Engineer in his knowledge gathering activity:

1. *the KNOWLEDGE PRINCIPLE*: "A system exhibits intelligent understanding and action at a high level of competence primarily because of the specific

knowledge it can be brought to bear: the concepts, facts, representations, methods, models and heuristics about its problem domain"
2. *the WELL-FORMEDNESS THRESHOLD*: "For each task, there is some minimum knowledge needed for one to even formulate it"
3. *the COMPETENCE THRESHOLD*: "Difficult tasks succumb nonlinearly to knowledge. There is an ever greater payoff (convenience) to adding each bit of knowledge, up to some level of competence. Beyond that, additional knowledge is useful but not frequently needed (e.g., it is used to handle exceptions or anomalies). When a point is reached in which almost all of these exceptions are handled as well, the addition of knowledge carries very little usefulness, and should be avoided"
4. *the INCREASING BREADTH/LIMITING DEPTH PRINCIPLE*: "Intelligent performance often requires problem solvers to fall back on increasingly general knowledge, and/or to draw analogies to specific knowledge from seemingly distant domains. When this happens, and some unexplained logical or physical similarities can be detected between two facts A and B pertaining to distant domains, it is more useful to *put these similarities to immediate use* than to try to analyse in depth the logical paths which lead to these similarities"

This Chapter deals with some of the available knowledge *acquisition* and *representation* methods. In particular, Section 4.4 gives a general outline of a block diagram for the translation of a design activity into an ES.

4.1 - General Knowledge Representation for Design Purposes

Selecting a suitable representation for the domain knowledge is one of the first problems encountered when building a KBS. There are some general principles which should guide this representation, though there is a considerable degree of disagreement among specialists in the field. The views presented here are therefore to be taken as a generalisation of the authors' personal conclusions in this subject.

I. "real" knowledge is hierarchical: so should its representation be;
II. the (common) language which is originally used to describe a specific piece of knowledge ought to be "translated" into the KB without forcing it into a more precise frame than that which was originally its own: for example, if the original information is sketchy or vague or dubious, care should be exercised to retain some sort of approximation, vagueness or uncertainty in the pertinent KB objects;

III. though not necessarily so, it is advisable to code separate production rules at each level of the knowledge hierarchy: these rules may and may not be the same at different levels;

IV. since in most engineering cases the Knowledge Base is a changing one, it is advisable to use a flexible domain-knowledge representation, consisting in a "network" of cross-referencing rules. Avoid "tabular" knowledge representations, because they are rigid and not easy to modify.

In practice, these guidelines are not difficult to implement, and their consistent application constitutes already a substantial step towards the successful compilation of an ES. They result in general "procedural rules" which form a sort of "syntax" of the implementation language: by analogy, the "grammar" of the language consists of a set of types under which the available knowledge can be catalogued: the list here below is not intended to be either complete or absolute, since it has been derived from collected application experiences, but it constitutes a rather complete "dictionary" of knowledge classification that covers most Thermal Systems Design problems. According to this list, both the conceptual (qualitative) and the physical (quantitative) knowledge needed for the implementation of a Expert Assistant for Thermal Systems Synthesis can be regrouped under the following types:

1. *Objects*: components, fluids, controls and equipment in general, human operators;
2. *Properties*: attributes of an Object. They can be qualitative ("component A is more apt than component B to the type of service described by design specification rule R") or quantitative ("maximum gas turbine inlet temperature T_{max} = 1300°K"), or both ("Deaerator extraction pump should be a booster for the main feedwater pump AND its net head ought to be higher than 0.1 MPa");
3. *Relationships*: (virtual) actions that directly or indirectly connect components or processes. They can be physical ("Boiler needs air, fuel and preheated water as inputs") or logical ("If the fuel is coal, then a check of its sulphur content is necessary to decide whether to consider a DeSOx unit or not"). Relationships can be further subdivided into:

3.1 - *Time relationships*: relationships that have time (in the physical sense) as an independent variable. Example: "The availability of a sufficient mass flow rate of external cooling water must be ascertained *before* considering the possibility of inserting a water-cooled condenser in a fossil-fuelled powerplant".

3.2.- *Cost relationships*: relationships that have cost (either in the monetary or exergetic sense) as an independent variable. Example: "The exergetic cost of the unit mass of natural gas is higher than the exergetic cost of the unit mass of pre-treated urban refuse".

3.3 - *Causal relationships*

4. *Aggregations*: super- or subsets of a system. For example, a plant based on a combined cycle has both the steam plant and the turbogas as its subsets; in turn, both the turbogas and the steam plant possess their components as subsets: in this case, we would call "aggregation level 0" the level at which the single components are considered and described; "aggregation level 1" the level at which the turbogas and the steam plant are considered, and "aggregation level 2" the level at which the entire process is the "black box".
5. *Engineering assumptions*: all rules, specifications, descriptions, principles, assumptions and "rules-of-thumb" which derive from the direct experience of Process- or Design engineers. They include all the heuristics usually applied in the specific design activity.
6. *Procedures*: norms, laws, regulations of any kind. It is useful to include in this type also all the "approximate", "vague" or "uncertain" descriptions of operative procedures ("If excessive vibrations are detected on a compressor shaft, stall-check procedure must be launched").
7. *Analogies*: all types of analogical reasoning known to apply to the field of application of the ES (see also Section 5.1 about Meta-structures).
8. *Behavioural rules*: all qualitative or quantitative information available on the operation of any component of process of interest. It is advisable that these rules also include approximate and uncertain behaviours.
9. *Approximate quantitative analysis*: "bulk" estimates or first-order approximations to process- or component calculations. Example: "If a gas turbine blade row is air-cooled, there would be no need of special alloys up to gas temperatures of about 1100°K"; or "Rough energy balance of the deaerator is: $m_{steam} * h_{steam} + m_{water} * c_{p,water} * T_{water} = m_{(condensed\ steam+deaerated\ water)} * c_{p,(condensed\ steam+deaerated\ water)} * T_{(condensed\ steam+deaerated\ water)}$"

Using this grammar, and including in the paradigms of the ES to be constructed the syntax described by rules I-IV above, a sufficiently detailed, well-organised, easily accessible knowledge representation is easily achieved.

4.2 - Knowledge Acquisition

Acquisition of knowledge is a formidable problem in its own. In 1977, the so-called *Feigenbaum bottleneck problem* was formulated that stated: "Expert knowledge acquisition is a problematic bottleneck in the construction of application-oriented Expert Systems" (as quoted by [20]); after over 20 years, it is still true that a good method for constructing and maintaining (updating) the KB is the best guarantee of success of the ES. A substantial portion of the bottleneck problem is constituted by the wrong use that is made of expert's advice: often, it is required of the Domain Expert to systematically analyse all implications of a certain design decision, thus completely ignoring a crucial fact, namely, that it is the application experience, and not system theory, the important content of the expert's Knowledge! So, in this sense, this part of the bottleneck is caused by a wrong approach on the part of the Knowledge Engineer, and can be easily removed. Nevertheless, the problem does not lie in *collecting* domain-knowledge (though this can be difficult in some specific cases, see [34,50]), but in *systematically storing it in the KB*. This can be best done with the help of specifically designed knowledge-acquisition computer codes, which "guide" the user through the process by making a large body of meta-level knowledge-acquisition rules accessible to him: unfortunately, general KB-acquisition codes are not readily available, since ES-assisted KB acquisition has been shown to work for rather small problems, but becomes unwieldy as the complexity of the domain increases. There are basically two "modes" this "expert knowledge acquisition" can be performed: the *automatic* and the *interactive* approach. In the interactive mode, the Knowledge Engineer would use his specific expertise in knowledge structures to help the designer fill out certain structures with portions of the available domain-specific knowledge. The general structures of knowledge remain the same from one application domain to another, but they are filled with different concepts and data (different Objects) according to the specific problem under consideration. The representation is bound to be closely tied to conditional rules, and a significant degree of code/user interaction is needed. In the automatic mode, the Knowledge Engineer is given the domain-knowledge in "chunks", i.e., several pieces of information at a time, and works on them independently from the user, trying to "induce" from these primitive facts the structure of the knowledge it will have to use. Adding other "chunks" helps him verify his conclusions, to amend them as necessary, and to generalise further the induced knowledge structure.

Irrespective whether knowledge is acquired automatically or interactively, the following general guidelines ought to be abided by when implementing a Knowledge Base:

- conclusions must be fed to the ES in the form of a finite set of discrete and specifiable elements;
- evidence (observed facts or data) about the problem must be reliably obtained by the programmer or by the system itself;
- initial assumptions must limit the problem to a highly specialised area, thereby directing the reasoning;
- it must be possible to use the KB to link *a posteriori* the evidence to the conclusions;
- the reasoning control strategy built into the production rules of the ES must in some way mimic the "human" expert one: for instance, when in doubt, a human expert asks for additional information before continuing in his solution strategy, or tries to rely on different data if certain other data are unavailable, or again tries a different path to the solution if a deadlock is apparent, and so on. Interactive prompting, widely branched decision trees, backtracking etc. are all techniques that help implement these features;

- an ES must possess another "human" quality: when duly prompted, it must be able to explain in detail (according to its own logic) all the steps which it undertook;
- if the required knowledge is nontrivial and does not exist in any form, the acquisition effort must include its generation and formulation.

For what the individual rules are concerned, a general framework in which they can fit is the so-called "Classification Model", which consists of:

- a list of possible conclusions
- a list of admissible data
- a list of rules relating data to conclusions

This classification is very general indeed: so, if the rules operate on the data according to a paradigm that tries to validate the admissible conclusion (goal-driven mode) by checking whether all data are contained in the premises (the LHS) of each goal, the Inferential Engine is of the BC type (Section 3.1.12). Conversely, if the rules scan the data to compare them with the premises of the goals, so that the goals can then be inferred, the Inferential Engine is of the FC type (Section 3.1.13).

Notice that a rule-based logic is not as omnicomprehensive as it may seem: considering a general production rule of the form:

IF (logical conditions are satisfied)
THEN (affirm the conclusion)

one may notice that it can well represent reasoning based on causal, temporal and functional relationships among objects, but cannot adapt to other widely used forms of reasoning like rules-of-thumb, heuristics, or similarity considerations. Thence, specific methods must be devised for simulating these "indirect" thinking modes. This can be done in several ways: our first choice is to attach a probabilistic value to the conclusion drawn by an ES. From the coding point of view, there is no problem in doing so, if the probabilities of each object in the KB are known. The problem has been found to lie in the knowledge gathering: experts tend to assign inconsistent values to probabilities of conflicting events (in that, for instance, implicitly deny the $1-p$ probability to an occurrence A even after they themselves have assigned the value p to its mutually exclusive event B). Heuristics and similarity considerations fall under the realm of qualitative knowledge, and can be handled by *ad hoc* paradigms (see Sections 3.5 and 3.8).

A meta-level problem is that of the *appropriateness* of the gathered knowledge. Though it is apparent that the nature of a piece of knowledge must be evaluated as well as its source, the ways in which this task can be accomplished are by no means clear and well-defined. It is always advisable to subject the gathered knowledge to a first-level screening by an additional (independent) Domain Expert, and it might be even necessary to iterate the process until a satisfactory level of appropriateness has been achieved. One should always keep in mind that, even if they came from a fully reliable source, noisy, uncertain and unreliable (non self-consistent) data will no doubt generate difficulties in the reasoning procedures, and will also affect the results in unpredictable ways, thus transferring in fact their unreliability to the conclusions drawn by the ES: this effect is called *logical error propagation.*

Quantifying the amount of data which has to be included in the KB can also be a problem: in general, knowledge-intensive and data-intensive representations have been found to be more successful in AI procedures: one may use as guidelines the so-called Knowledge Principles listed in Section 4.0.

All things considered, it has to be underscored that engineering reasoning is a *second level* reasoning: it subsumes, so to say, that more "elementary" reasoning capabilities are already present in the reasoning subject (language and image recognition, direct acknowledgement of obvious causal relationships, etc.). This is not a fine point: when simulating the reasoning process of an engineer, *it is logically acceptable to assume that his reasoning follows the normal rules of logic at least as formalised in a low-level, common language expression of social interactions.*

4.3 - KNOWLEDGE DECOMPOSITION: SEMANTIC NETWORKS

There is an AI formalism which allows for a radical decomposition of the Knowledge Base into pieces of information so general and yet so detailed that they can be properly called "information bits"; this formalism is called *semantic networking*, and offers many advantages over standard logical methods: a major advantage being that of allowing a very compact representation of the aggregation rules. The general description of this formalism has been given above (Section 3.4): as an example of its application, let us represent the fact "The turbine in the Schaffhausen plant is a Francis" in terms of a Semantic Network. There are several possibilities, one of which would be the following:

TURBINE → type → FRANCIS
PLANT → contains → COMPONENTS
TURBINE → included in → COMPONENTS
SCHAFFHAUSEN → is a → PLANT
FRANCIS → is a turbine in → SCHAFFHAUSEN

Here, "type", "contain", "included in", "is a" and "is a turbine in" are *operators*, while all terms in capital letters are *objects* contained in the Knowledge Base. Now, if we want to represent another fact (distinct from the previous one), like "The ten turbines in the Wujiang plant are all Kaplan", we should of course add some facts to the Knowledge Base, but will also have to modify the Semantic Network to allow for the two different plants (Schaffhausen and Wujiang) to be recognised as belonging to a common Class "plant", and the Francis and Kaplan turbines to be identified as distinct members of the Class "turbines". One possibility would be that of introducing the concepts of "INSTANCE OF A TURBINE" and "INSTANCE OF A PLANT":

INSTANCE OF A TURBINE1 → type → FRANCIS
INSTANCE OF A TURBINE2 → type → KAPLAN
....
INSTANCE OF A TURBINE11 → type → KAPLAN
PLANT → contains → COMPONENTS
TURBINE → included in → COMPONENTS
INSTANCE OF A PLANT1 → type → SCHAFFHAUSEN
INSTANCE OF A PLANT2 → type → WUJIANG
INSTANCE OF A TURBINE1 → is a → TURBINE
INSTANCE OF A TURBINE2 → is a → TURBINE
....
INSTANCE OF A TURBINE11 → is a → TURBINE
INSTANCE OF A TURBINE1 → is a turbine in → INSTANCE OF A PLANT1
INSTANCE OF A TURBINE2 → is a turbine in → INSTANCE OF A PLANT2
....
INSTANCE OF A TURBINE11 → is a turbine in → INSTANCE OF A PLANT2

The distinction among "types" (classes) and "individuals" (members) allows even for the proper representation of anomalies and exceptions (example: "one of the 10 Kaplan turbines is a bulb turbine"), because, if "instance no. 8" is a bulb turbine and "bulb turbine" is a (special kind of) Kaplan, the logical mechanism represented by the Semantic Network will accommodate the exception with no problems (without actually recognising it as an exception).

This brings up a final but not minor point: one of the characteristics of a Semantic Network is that of *not being able to check on its own congruency*: because it does not contain any "meta-rule" which can be invoked to scan the contents of its knowledge base singling out possible incongruencies. The only congruency which is assured is that *internal* to each relationship: it is precisely this lack of topological constraints which makes the network so flexible and powerful. In particular, inference rules can be varied independently from knowledge acquisition rules, and the resulting programming freedom can be of great advantage when handling large, realistic knowledge bases. There is a price to pay for this: congruency check, if needed, must be coded into the program, making the network somewhat complex.

The distinction between classes and members is not peculiar of semantic networks: there are several "Object Oriented languages", like Smalltalk, Quintus Prolog, C++, etc.(see Section 5.4) whose semantics are based precisely on this distinction.

CHAPTER 5

EXPERT SYSTEMS DEVELOPMENT TECHNIQUES

5.0 - ARCHITECTURE AND SOFTWARE

Engineers are accustomed to "traditional" numerical codes that consist of two sets of objects: Data Structures and Algorithms. Data Structures can be numerical, logical or "structural" (like the information about process connectivity, see Section 2.4.1.5), and are kept physically separate from the manipulating algorithms that constitute the actual procedural "solution" of the problem. The "execution" of a code consists of the algorithms performing certain pre-determined operations on the data in order to produce some "Results". Within a certain code or package, these algorithms are invariant, and the procedure that leads from the input of data to the output of results is entirely deterministic (see also Section 3.2). All the qualitative, and a great part of the quantitative Domain Knowledge is hard-coded, i.e., it has been translated into programming instructions, because the only quantitative Domain Knowledge that is present in the data is some information about design specifications, initial- and boundary conditions. By contrast, Knowledge-Based Systems ("KBS", see Sections 3.1.2, 3.1.3 and 3.1.4) are structured on a different model. They still consist of two sets of objects, Knowledge Base (KB) and Inference Engine (IE), which present the same interaction as that between algorithms and data, i.e., the IE "works on" the KB: but both objects perform very different functions from the "Data" and "Algorithm" portions of a deterministic code. For one thing, the Knowledge Base contains both the raw data and all necessary information to operate on these data (i.e., it contains both data and algorithms). This "necessary information" is more than just a collection of algorithmic procedures though, because it consists of rules (see Section 3.1.6) that may and may not be algorithmic: for instance, qualitative knowledge about analogies of certain portions of the KB with other portions of the same or a different KB may be expressed by a subset of these rules. The Inference Engine is a collection of rules that perform logical operations to achieve the Goal (= Result) specified by the user. In the direction of the hierarchy of knowledge described in Section 3.1.2, KBS are one step above Numerical Solvers, in the sense that they are one step more removed from being “non-intelligent tools”, and one step closer to being “imitations of human minds”.

The general problem of building an AI software is that of creating a set of tools for the construction, maintenance, manipulation and use of a possibly large, domain-specific KB. There is no universal procedure or standard set of guidelines to achieve these goals: to a large extent, the Data influence the Inference Engine, and vice-versa. However, two general principles should be adhered to while building an ES:

i) for each possible domain of application (more than one can be foreseen for most practical cases), it should be possible for a Domain Expert to "educate" the program interactively, checking and correcting, if necessary, its behaviour;
ii) it should be possible for a non computer-specialist Domain Expert to assemble, maintain and modify the Knowledge Base

Traditionally, the task of physically coding the AI constructs which the program consists of is the task of a human "assistant" (the Knowledge Engineer, see 3.1.17) who interacts with several Domain Experts and, being a computer specialist and a general AI practitioner, puts together the building blocks of the code using some of the well-known "structured programming" techniques. Lately, some attempts have been successfully made to substitute an "artificial assistant" for the human one [20,30]: so, in the following Section, we will talk about an "Assistant" without specifying whether human or machine.

The choice of the "language" to use is naturally fundamental in practice, but, from our point of view here, irrelevant: all available languages have basically the same set of logical procedures (though admittedly they may represent them by different symbols), so that if a code is originally written in, say, C++ and then we try to construct the same code in SMALLTALK, we will be faced with formal translation problems, but no substantial one. As a matter of fact, there are several automatic translators which can facilitate (or eliminate at all) this task. Once the language has been chosen, the assistant will have to gather from the expert the fundamental knowledge about the problem(s) to be solved, and decide *what* to implement in the ES and *how* to implement it. The first attempts towards the implementation of ES have been based on the so-called *machine-centred approach*: the ES had to be built in such a way that the machine would be in control, i.e., would prompt its users and instruct them on what to do in different phases of the problem solving. This approach is practically feasible only for very simple and logically limited tasks (i.e., for tasks that possess only one- or two levels of meta-knowledge and whose decision tree has not too many branches), but it suffers from another more severe limitation: it implies that *all* of the necessary knowledge has already been collected and correctly coded, that *all* of the possibly applicable rules have been properly implemented in the code, that exceptions have been handled as well, that analogies with other “similar” tasks have been correctly drawn and their consequences incorporated, etc.. This is both virtually impossible to achieve, and not correct from an AI point of view: it would make, is a sense, the ES “deterministic” and similar to a closed system, while KBS’s are by definition open-ended procedures. This type of paradigms is therefore now considered inappropriate, and has been abandoned in practice. Its replacement is the seemingly more ambitious, but in reality easier to implement, *human-centred approach*, which attempts to reproduce in the ES the (or “a”) mental model of a designer or an engineer. The underlying principle of this method (first discussed in [56]) is that of *functional decomposition*, i.e., of capturing the mental models of the user and putting this knowledge to use in the implementation of an ES that performs according to these models, and it is capable of interacting with the user to test a bit of knowledge it has retrieved, to communicate the results of its inference even at intermediate stages, and to acquire more information during its execution-time. For what concerns the Design of Thermal Systems, the human-centred approach has been described both in principle and in detail in Chapter 2. It is possible (and interesting!) at this point to put even the “Universal Design Procedure” (“UDP”) of Chapter 2 in a different perspective: let us examine the possibility of embedding it into a meta-level “Problem Solving” procedure. Consider Fig. 5.1, and notice that the (entirely qualitative) paradigm described by this logical flow chart contains all of the activities of the UDP: but what is striking is that the same paradigm can be used to describe the way we would for instance approach the problem of “choosing a route”, or “tying a string”, or “picking a book”. So, this Problem Solving Procedure (“PSP”) is at a meta-level with respect to the UDP, and it is the PSP that we should model (attempt to reproduce) in our Expert Systems!

A generally accepted paradigm for this stage of development of an ES, which mimics the mental procedures of an engineer, and that it must be considered by the Assistant, is the following:

1. what are the goals?
2. what are the premises?
3. what are the constraints?
4. what are the general laws governing the phenomena?
5. are these laws regular, or do they admit exceptions? Do these exceptions admit a systematisation of any sort?
6. are there intermediate checks which can be made during the solution procedure to avoid "running astray"?
7. can the basic knowledge about the problem be formulated in chunks? If yes, how well-structured are these chunks? Does each sub-domain (chunk) admit its own IE? Is there a single IE responsible for the management of the entire body of knowledge?
8. are there implications which explicitly or implicitly require access to a different KB? What kind of expert advice (in a different domain) should be sought after?

9. can the expert formulate a macro-code? if the Assistant formulates a different one, can the expert critically review it?
10. can the expert bring about one or more "test cases", for which the solution is known to exist and is -in qualitative and/or quantitative terms- entirely or partially known?

At this point, the Assistant can start writing the code. It is good practice to start worrying about the large-grain structure first, and then refining it level-by-level. The fundamental structure of an ES is depicted in Fig. 5.2: it consists in principle of only two parts: the inference engine “IE”, which is an ordered set of rules properly constructed so that they can draw inferences from a proper set of “antefacts”, and the knowledge base “KB”, which is a non necessarily ordered set of “facts” about the general and/or particular behaviour of the problem domain. In essence, the KB is ideally a logical container in which all the antefacts of the IE-rules are stored. In practice, since the KB does not need to be either self-consistent nor ordered, it does not necessarily contain all of them explicitly, and neither contains them at the level at which they are needed by the IE: so, some sub-sort of “secondary” or “auxiliary” IE may be contained in the KB, to help sort and organise the different chunks of information to make them suitable for the IE to work on them. It is apparent that such a skeleton is not a workable code: in general, we will also need (see Fig. 5.2) a Knowledge Acquisition Utility “KAU”, which collects (see Sections 3.1.2, 3.1.17, 3.1.18) the raw information bits and organises them into a KB proper; and a (Graphic) User Interface “GUI”, which allows the user to interact easily and with a high-level language with the ES. These two modules ought to be regarded as necessary parts of any ES. Additionally, it is highly advisable the ES possesses an interactive “On-Line User’s Guide” (“OLUG”) which can be consulted on-line by the user, to check on input/output procedures, to control the inference mechanism, to verify intermediate results, etc. Finally, there must be a workspace which contains the results of the inference, so that they can be accessed, modified, translated into additional knowledge for the KB, sorted, analysed in any way the user wishes (even by use of another external ES), and displayed in the form best suited to the user’s needs. Such a workspace is obviously a (Knowledge) Data Base, and it is called the Context [74] or the Working Memory.

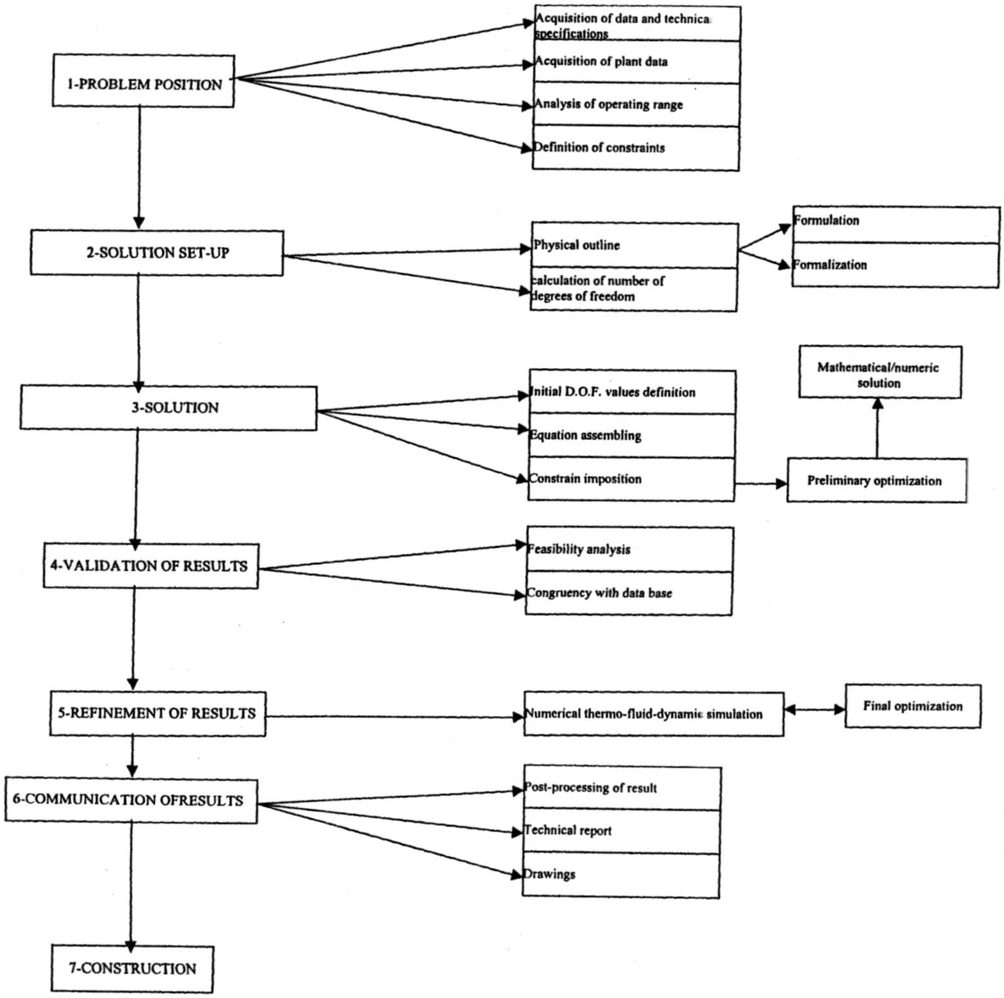

Figure 5-1. Logical scheme for Problem Solving Procedure

Once these modules have been defined, it is easy to start “filling up” the blanks, by writing modular sub-codes that perform the required actions for each module. As soon as the most elementary checks have been done (mostly, syntactic self-consistency and grammatical correctness), the code should immediately be made available to the Domain Expert(s), so that the code's "education" can be started. If a Domain Expert is not a computer expert, the Assistant will have to run the code for the expert and actively interact with him to complete the preliminary education of the code. Then, the Assistant should start working on the database managing portion of the code, to ascertain whether condition ii) above is fulfilled as well, i.e., whether the Domain Expert can assemble, modify or manipulate the Knowledge Base. In the entire process, a secondary goal of the Assistant is that of keeping the Domain Expert away from coding sorrows, so that his only engagement may be the definition and organisation of the data base, on the one side, and the definition of the proper logical structure of the Inference Engine, on the other: it is highly desirable that the expert is not bogged down by questions of programming and/or code control structures.

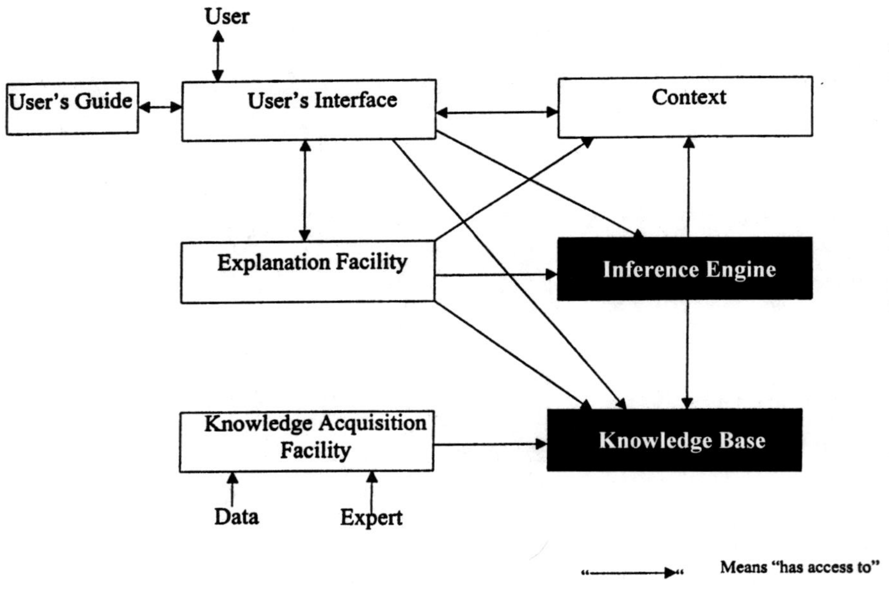

Figure 5-2. Structure of an Expert System

All in all, the process can be guided by good common sense. There are, as mentioned above, some Automatic Assistants (which of course are ES themselves); but their degree of success is variable and aleatory (what works well in medical diagnostic AI codes might not perform as well in apparently similar mechanical failure diagnostic codes), and there is up to date no substitute for a human intermediate between expert and computer. There are cases in which the Domain Expert is also "his own assistant", so to say, but, in today's highly specialised technical environment, this is rather the exception than the rule.

There is a number of general-purpose "shells" available today, which promise to perform well on any Knowledge Base they may be applied to. It is clear that the more general a shell (or whatever piece of software, for that matter) is, the more limited its fundamental set of logical rules will be. For this reason, "general" shells are apt to be used either in tutorial applications, to get a "feel" for AI techniques, or in logically simple problems, where they are called to operate on a not-too-fractionated knowledge base. Since really universal shells are not in sight, the potential user should instead search for a tool (an AI language) with the widest possible range of capabilities.

5.1 - Meta- and Macro-structuring

Sometimes it is possible to recognise *logically similar* patterns in seemingly different problems. For example, the procedure for locating the economic optimum of a component (say, a heat exchanger) and that for finding the minimum weight of a gas turbine engine are *logically identical* under many aspects, though of course the procedures they invoke are very different in type, complexity and character. If one examined *analytically* the two procedures, *each one in its own syntax*, very few analogies, if any, could be found. But, if the examination is repeated, *this time at a higher level of abstraction*, for instance in the syntactical context of the general optimisation procedure, similarities become striking. On the basis of these considerations, it can be concluded that it is very likely that, given two problems which display some degree of "common sense" similarity ("minimum cost" and "minimum weight" in the example above), it is possible to construct a higher level procedure which could contain (*embed*) the two separate procedures applicable to the problems under consideration. This (in all likeness, logically simpler) procedure is called a *meta-procedure*, and its implementation offers the obvious advantage of being able to solve two physically distinct problems. In some cases, there may even be a meta-meta-procedure, at an even higher level of abstraction: but, for most practical problems, the second level is high enough. It is also possible that, once a meta-procedure has been implemented, it is found capable of encompassing a third problem, "similar" to the previous two (for example, that of finding the minimum volume for a heat storage tank in a solar plant): a little thought reveals that there are many "similar" problems in engineering design, and that there are an even larger number of "almost identical" problems in plant maintenance and operation. Correspondingly, it will in general be possible to solve several *classes* of problems via the *same* ES, provided suitable features have been inserted in it for handling different knowledge bases. A meta-procedure is an example of logical *meta-structuring*; the process of collecting under a single ES the solution procedures for a large number of *diverse* problems is called *macro-structuring*. A "*macro*" is, in AI user terms, a group of procedures capable of handling (via the proper incorporation of lower-level specific sub-procedures) problems which apparently do not share the same Inference Engine (as a matter of fact, they may well have to be handled by different versions of a meta-engine).

5.2 - Qualitative Reasoning

It is quite obvious that to construct an ES, there must be some prior from of comprehension of the behaviour of the physical system (or class of systems) under examination. A critical review of the mental activities involved in the modelling of a system leads to the identification of the following general modelling paradigm:

1. creation of a general ideal representation of the physical interactions among the different parts of the system;
2. translation of this ideal representation into an explanation of the gross behaviour of the system (understanding why and how the system performs its tasks);
3. refinement of the general representation of the system by including the non-ideal discrepancies introduced by the reality of the components, by the non-ideality of the processes, by the non-linearity of the materials, etc.;
4. general understanding of the "asymptotic behaviour" of the system under limit conditions;
5. detailed reasoning about the qualitative links between the behaviour of the different sub-sets of the system and that of the system as a whole;
6. possibly some reasoning about the inverse design problem (given a certain description of the required performance, devise, characterise and "size" approximately the components that would produce such a performance);
7. formalisation of the general model and of the sub-models;
8. application and testing of the model.

In all this modelling activity, actually only steps 7 and 8 imply by force quantitative reasoning: in procedure just outlined, it is qualitative reasoning that is of importance! Therefore, it is mandatory that when assembling an ES, due care is taken in: a) creating the *right* model, and b) properly including in this model the qualitative knowledge about the real system.

Except for trivial cases, there is no definition of "a right model". Actually, history of science (and, more simply, our own experience!) shows us that models of the same physical phenomena may change according to technological knowledge and need[1]. For our purpose, it is necessary that a model be endowed with the following requisites [18]:

- the model structure and the system's physical structure must be *topologically equivalent*: no fundamental component or flux must be left out;
- all of the fundamental interactions (connections) among the various sub-sets that compose the system must be *explicitly* considered in the model;
- all the interactions of the system with its immediate surroundings must be *explicitly* included in the model;
- if specific characteristics of the system are known in advance, then the model must be able to reproduce them at least approximately. This applies for instance to steady-state performance (e.g.: the model must have the same efficiency and generate the same power as the system), or to limit cases (e.g.: if it is known that the system behaves in a certain way during start-up, the model must reproduce at least qualitatively this behaviour);
- there must be an order-of-magnitude match between the relevant parameters of the model and those of the system (e.g.: the mass flow rate of air at compressor inlet in commercial turbogas sets, m_1, varies between 3.5 and 4.5 kg per kW of net generated power. A "Gas Turbine Model" must assume or produce values for m_1 that remain within this range).

The logical relationships between a system and its qualitative and quantitative models are depicted in Fig. 5.3. There is of course a direct analogy between the real behaviour of the real system, the quantitative behaviour of the quantitative model and the qualitative behaviour of the qualitative model: this is a *logical* analogy, and is represented in Fig. 5.3 by the horizontal arrows. But there must be another form of analogy between the qualitative and the quantitative behaviour and the qualitative and quantitative models: this is a *functional* analogy, in the sense that both the "general" and the "limit" form of the descriptive equations must match in shape and order-of-magnitude. Since it is expected that the quantitative model correctly describes the real behaviour, this means that the functional analogy can be extended from the qualitative to the real behaviour as well. This "model of a model" represented in Fig. 5.3 views the qualitative model as an *abstraction of the quantitative model rather than of the real system*. This may seem a paradox, but it is not so: if we reflect upon the mental procedure we use when describing a qualitative model of, say, a combustion chamber, we see that our qualitative model ("gas is injected and

[1] Actually, scientific models have been subjected in the past (and still are today, though to a lesser degree!) to direct and indirect social-cultural censorship: probably the most (in)famous example is Galileo's persecution dictated by dogmatic intolerance of the Catholic Church, but modern forms of censorship abound. Consider for instance the practical banning of cognitive science (and of its models) by the East European governments in the 30's and through the 60's; or the resistance offered today by a large majority of a biased public opinion in Western Countries towards "self sustainability" theories, that push for the adoption of rather drastic measures for a real rationalisation of the final use of energy resources...

mixed with air in certain proportions, combustion sets in and the temperature of the mixture is raised almost isobarically to a limit computable by an energy balance, then the hot combustion products are ejected from the chamber;

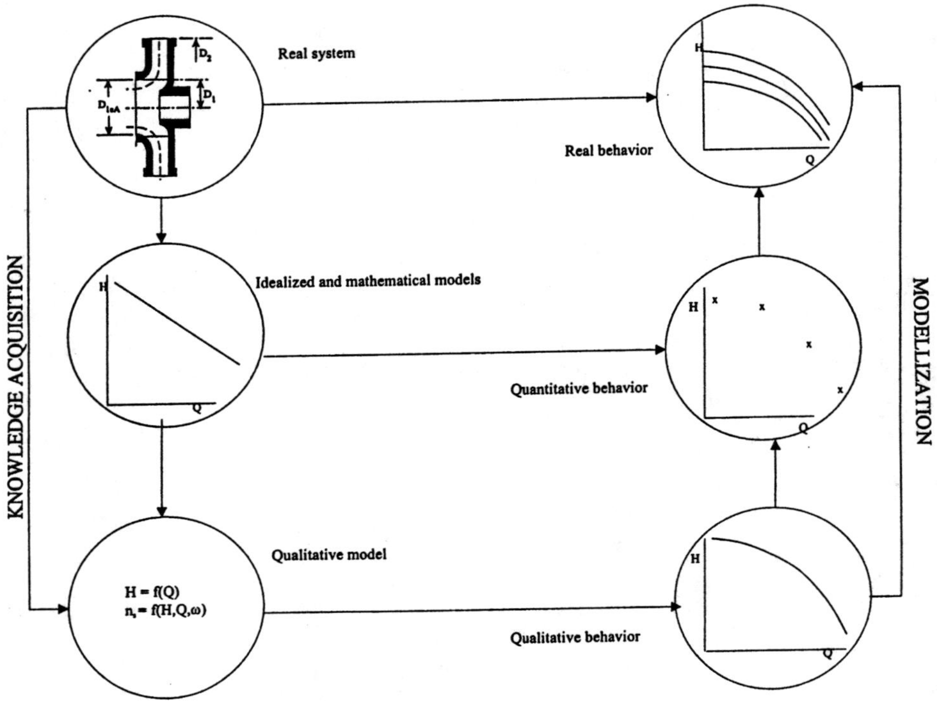

Figure 5-3. Relationship between system and its models

thermal losses are usually low, and are due to thermal conduction/convection through the walls of the chamber and to incomplete combustion") relies on some quantitative information that we unconsciously draw upon. The "certain proportions" of the mixing represent a qualitative reminder of the stoichiometric ratio; the "almost isobaric" combustion has been (correctly!) derived from our quantitative knowledge of the losses in flow of a fluid through a wide cylindrical chamber; the "low thermal losses" and "incomplete combustion" are qualitative expressions of the exact heat transfer calculations and of the chemical calculation of the real combustion reactions that we have either performed ourselves or have quantitative knowledge of. As a consequence of this "subconscious priority" that quantitative reasoning enjoys, the representation of qualitative behaviour and the creation of qualitative models can be derived from the well-known principles of quantitative modelling, taking into account some not-so-obvious differences. In fact, the semantic of quantitative computation is so inherent to our education that we do not question signs ("-" and "+"), zero values, inequalities, intervals ("$3.5 < x < 4.5$"), shapes ("every continuous P^3 possesses a P^2 derivative"): better said, we neither question the concepts that are behind these symbols, nor have any difficulty in comprehending their interrelations and put them to work, stating for example that "IF ($3.5<m_1<4.5$) AND ($0.01\leq\alpha\leq0.02$) AND ($m_f=\alpha m_1$) THEN ($0.035\leq m_f \leq0.09$)". But qualitative reasoning is something different: actually, there is no exact qualitative equivalent of the above statement, because all what matters in qualitative reasoning is the sign and the order of magnitude ("OM) of a variable: so the most we can say is "IF ($OM\langle m_1\rangle = 4$) AND ($OM\langle\alpha\rangle = 0.015$) THEN ($OM\langle m_f\rangle = 0.06$)". A "Qualitative Mathematics" has been developed, which defines the four basic operations, derivatives, equations, tables of influence and so on: its description is beyond our scope here, and the interested reader is referred to [33,74]. What we can say is that Qualitative Mathematics can be put to use to locate possible "states" of a system (a *state* being defined as a situation in which all the values of the qualitative variables are known to assume either a positive or a negative value or otherwise be equal to 0) and the trends of its behaviour (that is, the sign of the derivatives of all variables of interest in that state). In addition, Order-of-Magnitude reasoning can operate on a set of variables not only with its usual operators "equal to", "larger than" and "smaller than", but also with a class of extended inequalities like "slightly greater than" or "much smaller than", which can be combined to generate a very fine mesh of inequality relations (like "moderately less than or approximately equal to" or "slightly or moderately larger than"), which can be used to define the intervals of interest for the relevant variables. There are packages [29,40] specifically designed to perform OM calculations.

A different and possibly more physical approach to qualitative reasoning is that of going back to the structure of the system to be modelled, and consider that a system operates to modify a physical situation by executing a process [10]. A physical situation of interest for a Thermal System consists of *Objects* (components and fluxes), their *Relationships* (mass- and energy exchanges), and *Processes* (transformations enacted by some of the components on some of the fluxes). Processes produce modifications of both the physical and the qualitative properties of Objects. This is the so-called Structural Operators theory, and its simplicity and power are the main reasons behind the success of Object-Oriented Languages, that are environments in which this theory finds its easiest and most direct application.

For the introductory but operative character of this book, we find that an operative model description would better suit our purposes: this description, which must be of immediate use in our applications is -in a logical sense-somewhere in between Structural Operators theory and Abstract Model theory. Consider the scheme of a general "Modelling and Simulation" activity, depicted in figure 5.4: there is only one intermediate step between raw data (Primitive Facts) and results (Derived Facts), and this step is called "Compiled Knowledge". It contains both deterministic and intuitive knowledge, both algorithmic and non-algorithmic operators: how are these logically diverse chunks of high-level knowledge produced? In the mind of the Designer, they are no doubt generated by *concurrent* application of both qualitative and quantitative reasoning to the same Knowledge Base. The entire process is regulated by a "Context Controller" that inhibits logical trespassing. We find schemes like those of Fig. 5.4 very useful, and suggest to keep them in mind when devising and assembling an ES.

The general Inference Motor of this Qualitative Process Theory proceeds as follows:

1. General identification phase: scan the KB, and on the basis of the existing Objects and of the applicable Process theory (which is domain-specific and ought to be imparted by the user) decide which Processes can exist in the current universe;
2. Specific identification phase: scan the KB an identify all active Processes on the basis of the (logical or physical) existence of their necessary premises;
3. Locally predictive phase: determine al changes that will be produced by each active Process. Consider cumulative effects: if several Processes act concurrently on a single variable, determine the net change in the value of that variable caused by the combined action of the different Processes;
4. Globally predictive phase: detect possible variations in the active/inactive status of all Processes of interest due to the changes caused by the currently active Processes.
5. Go to 1

It is apparent that this Qualitative Process Theory is in close analogy with both iterative solution methods of non-linear problems and with time-marching techniques employed in fluid dynamic simulations. As a matter of fact, though a process is by definition dynamic in its nature, this theory can be applied as well to "steady" processes, in which case it can iteratively reproduce the system's final response to modified boundary conditions. It is necessary to recall that we are describing a qualitative theory, and therefore "time" here is identified by its qualitative value "the time interval of interest", and three relational operators: "before", "after" and "during" (OM-type operators could be introduced here too, so one could talk of "long before" and "shortly after", etc.). So, the goal of Qualitative Process Theory is that of reproducing the approximate behaviour of the system by describing either its evolution in time via a qualitative response curve, or its steady-state behaviour by qualitatively describing its operating point.

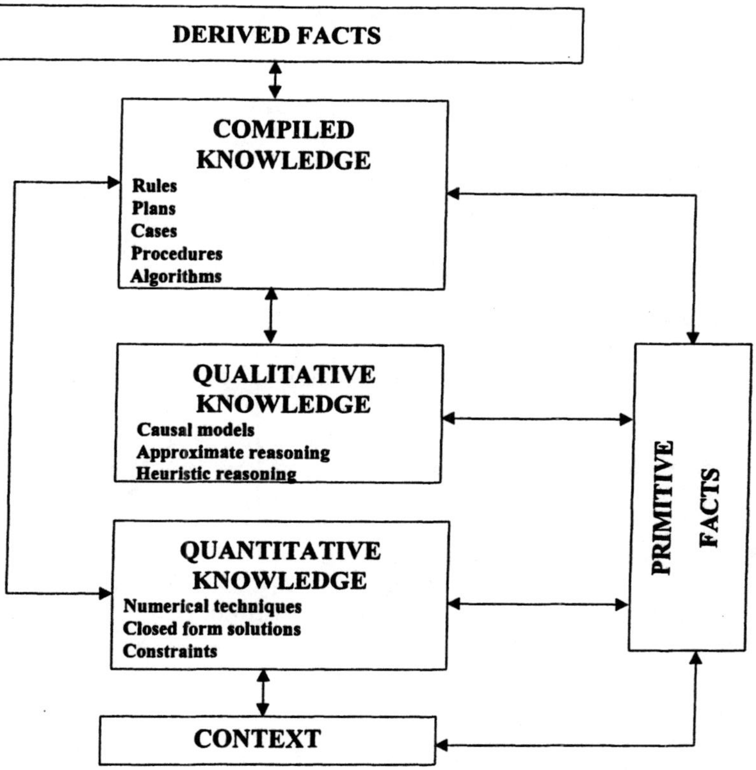

Figure 5-4. General scheme of a modelling and simulation activity

5.3 - Steps in the Development of an Expert System

5.3.1 - Motivation

There may be many reasons behind the decision to commit material and human resources to the development and implementation of an ES (see also above, Sect. 1.2): the most common ones are the need of formalising and clarifying a batch of knowledge which can be gained by convincing a human expert to make his reasoning explicit, and the need or the possibility of combining the expertise which can be gained from several human experts into a "body of knowledge" which can then be studied for consistency and reliability. There are, of course, lower level needs, so to say: for instance, operator overloading, or shortage of specification engineers, but they can be in the limit reconduced to one of the two main motivations stated above.

5.3.2 - STAGES OF ES BUILDING

It is important to start with small systems (both in the size of their KB and in that of the IE). Empirical validation is mandatory, and can be carried out in stages, as the system progresses towards its final form. Various stages in the development of an ES can be abstracted:

1- **Initial Knowledge Base design phase**: this comprises three principal sub-phases:

  1.1- **Problem definition**: the specification of goals, constraints, resources, participants and their roles;

  1.2- **Conceptualisation**: the detailed description of the problem; its possible breakdown into sub-problems, the formulation of each of these sub-problems (hypotheses, data, intermediate reasoning concepts); study of the possible correlations between the chosen formulation and its feasible computer implementation;

  1.3- **Computer representation of the problem**: specific choice of representation for the elements identified in the conceptualisation phase; possible choice of an existing general-purpose shell; choice of computer language; choice of user interface, choice of the operating system, etc.

2- **Prototyping**: the main purpose of the implementation of a prototype ES is to ensure that the project is feasible and to gain some idea of the ultimate costs and of the benefits that can be expected from the code. The prototype should give a feel of how the ultimate system will work, possibly offering also a realistic user interface. Building a prototype will take a non-negligible portion of the total time allotted for the project, and therefore it is important to solve possible fundamental problem-related issues *before* starting the prototype phase. A prototype will probably use some programming short-cuts to reach a "working" condition in which it can perform almost like its final successor, albeit in a limited fashion. It is very important to choose a proper subset of the general problem which will be dealt with by the final program: the chosen subset must include a representative sample of the knowledge typical of the overall model, but it must of course involve sub-tasks which are sufficiently simple to test. Also, the initial subset must be expandable in steps to allow for gradual prototyping, and must be such that this expansion reaches in the limit, without too large of a leap, the final problem formulation. Prototyping will reveal the problems that can be associated with capturing the domain knowledge, will quickly bring to surface any logical problem affecting the inference procedure, and will also show what kind of difficulties the prospective user will be faced with when using the final product. Usually (but not necessarily) the prototype will incorporate simple interfaces and use a somewhat rudimentary KB. It is advisable to have the final (prospective) users work on the prototype, so that they can understand it, criticise it, and in final effect, express their anticipated assessment of the ES. It should be the rule *that a code which does not satisfy either the needs or the standards of its final users ought not to go into production;*

3- **Refinement and generalisation of the Knowledge Base**: this is the phase in which, through iterative testing and painstaking double-checking, the code is adapted so that it can perform well in *all foreseen circumstances*; of course, this can not always be assured on a 100% basis, but the number of diverse occurrences is in general finite (and seemingly "infinite cases", like "for continuous variation of the pressure between 5 and 22 MPa", can be reconduced to a finite number of instances by, for example, subdividing the pressure interval in a number *n* of sub-intervals), and the general trends of the solution can often be inferred with the help of the ES itself. In any case, this is by far the most expensive phase (in terms of manpower), and can easily occupy a well-trained and experienced design team for months.

4- **Fielding**: in its "final" version, the code is installed on the site of its final use. As it is all too well known to senior and junior computer specialists (and also to anyone who has ever been on the user's side of a newly developed system!) almost no instance is known of a code which performs as it should when it is first mounted. Aside from system problems, interface incompatibilities with other codes and obvious bugs, for an ES there exists the real possibility of a logical fault or incompatibility with the *real* data of the real problem. The more extensive and careful the prototyping phase has been, the easier will be to solve substantial installation problems (we consider here "trivial" all "system" and "compatibility" problems, because we compare them here to possible faults in the IE or in the KB representation!). However, it is good planning to consider the possibility of having to "assist" the user in this installation, not only with System Specialists, but also with Domain Experts and Knowledge Engineers.

5.4 - AI LANGUAGES

Currently, one can find AI applications written in several programming languages; it is convenient to group them into three categories:

1. General purpose programming languages

These are standard high-level languages, like LISP or C, that can be extended somewhat to implement some of the utilities required by AI reasoning. Notice that even FORTRAN can be used to implement logical procedures, though its rudimentary logic functions make such a programming extremely awkward.

2. Languages based on a specific programming paradigm

These are languages specifically developed for AI programming, and each one is based on a particular formalism for knowledge representation: so, there are rule-based (like OPS5), frame-based (like KEE), network-based (like NEURAL), logic based (PROLOG), and the recently introduced object-oriented languages (SMALLTALK, C++)

3. Hybrid languages

These are languages that allow for the implementation of different programming formalisms. Better called "Integrated Programming Environments", they are in effect ES frameworks, or shells.

Some Authors consider a fourth category, namely:

4. Domain independent ES Frameworks (shells)

In practice, these are well-developed Hybrid Programming environments, usually endowed with sophisticated user interfaces. The distinction between a "shell" and a "hybrid language" is becoming of ever lesser importance: we shall refer to all of the "multi-frame" environments under the name of "shell"

As repeatedly mentioned in the previous Chapters, the typical task of an AI code is that of scanning a Knowledge Base to infer "conclusions" from it. The nature of the domain representation and the implied symbolic structures require programming languages endowed with special features that enable the user to implement the fact-rules relationships into a goal search. One of the most powerful features in knowledge representation is the possibility of forming higher level abstractions from the domain description: it is the ability to exploit this possibility that gives an AI code the "skill" of understanding and managing complex domains; the knowledge must be organised -so to say- by the program at execution time. Such a capability demands for entirely new programming paradigms, that must be now equivalent to a complete logical system (see Section 3.3) and must allow for a convenient practical implementation of logical rules[2]. A brief reflection will show that neither of these attributes pertains to traditional procedural programming languages, which were not originally devised to represent a coherent logical system, and are structured in such a way that the construction of a complete set of logical rules becomes very cumbersome. It is clearly impossible except for the simplest cases to construct for instance a logical inference FORTRAN procedure that can infer correct results from the asserted data, because the only "logical" constructs available are "IF...THEN...ELSE" and "TRUE " or "FALSE". In addition, often our Engineering tasks are equivalent to incomplete or ill-posed problems, and it is necessary to introduce common sense and heuristic representation of the problem domain at different levels of abstraction in a structured hierarchical framework:

[2] One of the most influential early researchers of AI, co-author of PROLOG and considered to be one of the forefathers of AI programming, Robert Kowalski, expressed this concept concisely in his famous definition "Algorithm = Logic + Control", meaning that a logic programming language expresses only the logic component of algorithms, while the controlling function is perfomed by the programmer [32].

all functions that are totally absent from the usual "deterministic" programming languages.

The dominant languages for AI programming are PROLOG, LISP and Object-Oriented Languages (SMALLTALK, C++).

5.4.1 - PROLOG

PROLOG (PROgrammation en LOGique) was developed in 1972 as part of a project in natural language understanding. It is the best known example of a logic programming language, and its use has been so widespread among AI-researchers that it can be properly said that PROLOG is not only a language, but rather a new way of "doing logic" [20]. The original programming paradigm was based on ALGOL, a now obsolete early high-level language, but soon original low-level procedures were developed and implemented. PROLOG uses first order predicate calculus, and thus it can assert something about its logical variables (objects in the domain), but not about its own predicates.

Predicate calculus is based on descriptions of relationships between two entities, for example pump(Fluid_type, Flow_rate), pump(Fluid_type, Pressure). Through inference rules it is possible to compare and associate elements and/or variables of different predicates and infer deductions. A PROLOG program is a set of facts and rules. A *fact* is written as a *predicate*, like:

produces(gas_turbine, power)
needs(combustion_chamber, fuel)
needs(combustion_chamber, compressed_air)

Once a list of facts has been set up, it is possible to query the data base to answer user questions, like:

?- needs(combustion_chamber, fuel)

(The answer will be YES)

It is possible to bind a variable to an element of a predicate. In this case PROLOG will return all the terms that satisfy the query:

?- needs(combustion_chamber, X)

X= fuel
X= compressed_air

In the query *needs(combustion_chamber, X)* the variable X is not instantiated. The program searches the database looking for a fact that satisfies the question. The fact *needs(combustion_chamber, fuel)* matches the query and the variable X is therefore instantiated to fuel (*X=fuel*). Then PROLOG will also find the fact *needs(combustion_chamber, compressed_air)* which matches the query and the variable X will be instantiated to compressed air *(X= compressed_air).*

The procedure is clear: PROLOG will search the entire available database to find all of the possible instantiations for any variable X, Y, etc. contained in the query. A limit of this representation is that PROLOG assumes that the whole domain is represented in the facts list. If a possible solution is not included in the fact list that solution will not be found: notice that, if the Knowledge Base is updated by adding that particular solution to it, and the program is run again, then the "new" solution space will contain the new solution just added: PROLOG's logic is non-monotonic (adding new rules or facts to the data base modifies the output) .

If we want to know whether the Object 'combustion_chamber' needs both air and fuel, PROLOG uses the predicate AND (represented by the symbol ",") to link the two separate queries:

?- needs(combustion_chamber, compressed_air),
needs(combustion_chamber, fuel)

(The answer is YES)

Suppose now we want to find whether there are Objects that need the same input. In order to describe this example we must introduce new facts:

needs(compressor, compressed_air)
needs(pump, water)

The query

?- needs(combustion_chamber, X), needs(compressor, X)

prompts the program to look for an Object which both combustion chamber and compressor need as an input. PROLOG searches the database and, since the first goal is satisfied by

needs(combustion_chamber, fuel)

X is instantiated to fuel.

To satisfy the second goal, PROLOG now tries to find a fact which asserts that "compressor needs fuel". This search obviously fails, so the program deletes from its working memory the instantiation *X=fuel* . Since not the whole database has been searched there might be some facts that satisfy both goals: PROLOG "backtracks" to the first Object in its Knowledge Base and resumes the search. The program tries again to satisfy the first goal, and finds that air is needed in a combustion chamber, so that it instantiates X to *compressed_air*; PROLOG tries again to satisfy the second goal. This time, also the second goal succeeds: X is definitively instantiated to *compressed_air* and the solution displayed.[3]

5.4.1.1 - The rules

PROLOG uses rules to describe relationships between logically related facts. Suppose we want to find all the possible connections between two adjacent Objects. This may be represented by the rule:

connect(X,Y) :- output(Z,X),input(Z,Y)

This expression can be read as: connect X to Y if the output Z of X is the same as the input Z of Y. PROLOG is endowed with backtracking capabilities that it may apply also in rules search to find the facts that satisfy the goal.

Another powerful search mechanism in PROLOG is recursion, which is an elegant algorithm that can be used to scan lists. Lists are ordered sets of Objects which can be considered as the union of the first element (the head) and the tail (the remaining part of the list). To test whether an element is a member of a list, the function (*connect* in the above example) calls itself until either the terminating condition is satisfied or the entire list has been scanned. For example, to check whether the element X is part of a list PROLOG begins by testing whether X and the head of the list are equal:

member(X,[X|T])

(The head of the list is separated from the tail by the bar operator |). If X is not equal to the first element of the list then PROLOG checks recursively if X is equal to an element of the tail:

member(X[Y|T]) :- member (X,T)

Consider the following example:

?- member(c,[a,b,c]).

The element *c* is not equal to the head of the list, *a*. So the recursive function is called:

X=c, Y=a, T=[b,c]
?- member(c,[b,c])

[3] Notice a direct application of qualitative thinking: "compressed_air" and "air" are assumed to be the same Object. The quantitative parameter (the pressure ratio) is neglected in a logical sense.

Again, the head of the list '*b*', is not equal to the element *c* and the recursive function is called:

X=c, Y=b, T=[c]
?- member(c,[c])

This time the element *c* is equal to the "head" of the list, and the search is terminated successfully.

PROLOG is a very powerful language, and represents as well a sort of "logic benchmark" that can be useful for tutorial purposes. It is to AI programming in a sense what BASIC has been to deterministic programming: though now somewhat outdated, it is still essential to a correct understanding of the more advanced programming paradigms like C++, SMALLTALK and the like.

5.4.2 - LISP

LISP is a procedural language, in the sense that a LISP program describes how to perform an algorithm: but unlike traditional procedural languages like FORTRAN, LISP is *functional*, because its syntax is derived from algorithms based on recursive functions. In LISP, programs and data are represented as symbolic expressions; Objects are *atoms*, a sequence of Objects (atoms) is a *list.* A list can contain other lists, and this introduces a built-in recursivity into the syntax. A list can be evaluated numerically (if its contents are numerical Objects and operators), in which case LISP performs just as an arithmetic calculator. But, symbolic lists may be evaluated as well, and the evaluation is carried out by first instantiating all components in the list except for the first one, and then evaluating the function represented by the first component as if it was applied to the set of (numerical or logical) values of the arguments. The user can create new functions using the language built-in functions; these new functions can be called at execution time and return the results of the operation executed on the functions arguments. Though in a more cumbersome and low-level fashion, LISP can execute the same operations on lists (classes!) as those described in PROLOG, including recursion and depth- and breadth-first searches. It must be remarked that it was in some LISP dialects that the first object-like lists made their first appearance, before SMALLTALK and other Object-Oriented languages were developed: this signifies that the functionally-oriented LIPS architecture, though over 40 years old, is in no way obsolete. Since it was the most popular "pseudo-symbolic" language at the time of the early work in AI, many programs were implemented directly in LISP. This fact had a profound influence on later developments, because it is still the case that many modern shells are implemented in LISP, which has become, so to say, a sort of Assembler language for AI applications.

5.4.3 - SMALLTALK

The Smalltalk-80 Programming environment was the first fully Object-Oriented Programming ("OOP" in the following) language, developed in the early 70s at Xerox Research Centre in Palo Alto. It can be in fact considered an extendible (or "open") language, meaning that an application code developed in Smalltalk becomes part of Smalltalk itself and can be used for further applications: in a very real and concrete sense, user-added utilities expand the language. The Object-Oriented nature requires a proper programming approach: instead of creating a procedure that controls program execution (like in procedural languages), the user creates an Object and a set of Methods (functions) associated to the Object that operatively describe how that Object interacts with others. Messages exchanged among Objects modify the factual or procedural contents of the Object that receives them: it is worth mentioning that the result produced within the receiving Object is an Object itself. Classes can be defined as groups of Objects sharing common properties. For example, numbers are Objects and each number, as an instance of the Class "number", responds to messages such as '+', '-', etc.. Classes are organised hierarchically so that each one automatically inherits methods from its Superclass. The most powerful characteristic of Smalltalk is probably the possibility of developing logically modular codes: it is possible to write applications by adapting pieces of code already available that incorporate the functionality of existing programs. This allows for significant reductions in prototyping times. From the operative point of view, all user/language interactions are handled by a Smalltalk Programming Interface, which is the Browser window where all the available Classes and Methods are presented in individual sub-panes. If a Class is selected in the Browser, all code segments written thereafter will refer to that Class: and since instances must abide by the constraints applicable to their Class, the syntax in effect directs the user towards a structured formalism. The Smalltalk Browser can best be described as a bookshelf partitioned by Types, Categories, Classes, Class or Instance selection, Message categories, and Messages. Actually, the Browser is a real logic tool which segments the Problem Domain (any Problem Domain!) into conceptual pieces, while at the same time guiding the programmer through the logically appropriate partitioning of the problem space. It has been said that "the expert Smalltalk programmer uses the Browser to transform himself into an Object and view the world through the various protocols", and one must acknowledge that this somewhat picturesque view of the user/language interaction is not exaggerated. In practice, the user first selects the Superclass Category in the Browser which represent the properties of the Object to be constructed; then, he selects a Sub-class to enter the Class partition, and clicks the Instance button to enter the description of an Instance of the selected Sub-class. Next, he selects the Category to instantiate the Object, which will at this point directly react by issuing a message consisting either in a result or in a set of instructions that can be sent to other Objects in other Classes which "understand" the received message: this procedure is called "bell-ringing". One can use the Browser to navigate the Smalltalk universe and find the right Class for the application of interest. This may represent a difficulty for beginners, because the Smalltalk universe is vast – it comes with all the Classes necessary to describe the computational world – and gets larger and larger as users add their own applications which, as said, become themselves part of the universe and can be used for further applications as new partitions of problem domains. Of course, one can do real OOP in Smalltalk by using the Browser as described above and be able to use it only if he possesses some notions of predicate and relational calculus. To facilitate getting hands-on experience, a powerful tool to debug and to trace the program execution, called the Inspector, is available. By using the Inspector, it is possible to correct programming errors directly during the debugging session, so that the changes will be saved in the code written in the Browser: there is no need to rewrite code lines. This is maybe the best example of the philosophy on which Smalltalk was constructed, and that is a very careful blend of user control and self-checking modes, embedded in a fully integrated superstructure and presented in an extreme user-friendly fashion.

5.4.4 - C++

C++ emerged as a development of the widely adopted C language, and it was designed with two main purposes: first, the embedding of the OOP philosophy in a C-based frame, second, the overcoming of the difficulties encountered in all problems that require the handling of large data- and knowledge sets. In such cases, trying to deal with data representation and variables organisation using procedural programming languages like C, FORTRAN, Prolog, Lisp, can become a very hard task. C++ possesses "Classes" that can be considered similar to the "structure" construction in C: they inherit capabilities, and allow for a correct identification of an Object and all of its attributes. C++, unlike Smalltalk, cannot be considered a full OOP environment though, and neither it constitutes an extendible language: it is not a Programming Environment where parts of existing codes can be used, adapted to the new problem and become themselves part of the Programming Environment. Nevertheless, C++ is becoming more and more popular among AI researchers, thanks to its portability, good compatibility with other applications, ease of use and powerful memory management features. Some of the constructions common to C and C++ are: the structure (for instance, the program starting point is defined by the same *function main*), the loop operators (*for...do* or *while... do*) and the *if-then* primitives, memory address handling, and so on. The compilation process goes through the following steps: C++ source code → C++ translation to ANSI C → compilation → Object code generation, and this shows in fact that C++ is no more than a *dialect* of C. In other words, there is no added functionality in the language with respect to what already has been implemented in C. The improvements of C++ with respect to C consist in a more elegant and easier to read code listing and in the introduction of "ranking", i.e., of the possibility of creating hierarchical structures in which Objects with the same common general properties can be arranged. In C++, as in C, one of the most powerful tools is the capability of *dynamic memory allocation*, by which Objects can be allocated to and removed from memory slots at execution time: creating a memory allocation for an Object requires the creation of a pointer to the initial position (address) of the allocated memory. The Object "pointer" is defined as a member of the Class "Pointers", and can be then used to point to (= refer the program to) a variable or other Objects. *Recursion* is another syntactical feature available in C++ which has been inherited from C: a function keeps calling itself recursively to produce a new output for an input which is the output of the previous calculation. The recursion process goes on until a terminating condition is satisfied or the entire data base has been scanned.

A Class can be looked at as an interface which specifies the variables belonging to it and the behaviour (methods) common to the Objects it contains. A C++ Class is defined by its name, by the Objects (data, variables, etc.) belonging to it, by the functions (the set of operations that may be applied to its Objects), and by the access level (public, private, protected). Multiple inheritance is also supported: that is, a Class can have more than one Superclass. This situation can induce ambiguities, because a Class might inherit conflicting features from different Superclasses: in these cases, either logical exclusion or priority-handling methods must be enforced by the user to resolve the conflicts.

Three important data structures for Expert Systems applications are lists, trees, and graphs: C++ is endowed with built-in search procedures to efficiently handle these structures. In C++, Objects are *per se* arranged in a vertical hierarchy, and this is a little at odds with the concept of OOP (in Smalltalk, Objects can be thought of as "file folders" spread over the user's desk and the user has to look for the right folder for the specific piece of code). This "top-to-bottom ordering" means that the nesting (tree- or list- or graph scanning) will be *in depth*, which makes the execution somewhat slower for large systems. Any different search method must be "called" as a function, and this creates some overhead: in Smalltalk, the message-passing protocol circumvents the problem. For the execution of a single loop, C++ will definitely be faster than Smalltalk, because its built-in in-depth expansion is inherently faster than message-passing. But for iterative searches, when several hundreds of scans are executed, the combined overhead due to subroutine linking and register restoring can be much more significant than Smalltalk's repetitive message-passing overhead. Furthermore, when Smalltalk Objects are executed once, they carry the entire overhead connected to the search method they enact, because the iterations are handled entirely within the Object itself, thus enhancing execution speed. In conclusion, while the structured protocols of C++ are more familiar to engineers, and thus may help reduce the amount of human resources devoted to programming, they do not necessarily generate "faster" executables: in extreme cases, a comparative assessment ought to be made on a prototype.

5.5 - Intelligent Shells

An Expert System shell provides an *AI programming environment* which includes an Inference Engine that will act on the user-given data according the rules stored in the knowledge base to develop an appropriate answer. The search engine can use separate Knowledge Bases in accordance to the specific problem domain. The advantage of using a shell instead of a programming language is the capability of developing a prototype in short time, because it is generally easier to learn how to move within a shell than it is to learn how to use a programming language. However, since the production rules and the heuristics are already implemented, sometimes specific facets of the problem cannot be perfectly represented in the knowledge base.

In selecting an AI shell, it is of paramount importance to investigate about the performance of some specific functions. The *Knowledge Base representation* should be clear and easy to implement. A *graphical representation* of the data base and production rules interconnection should be available, so that the correctness of the domain representation can be checked at any time. The *transfer of knowledge* from one representation to another should be relatively transparent to the user: this means that the shell should support a variety of data types (numerical, symbolic, etc). The *Inference Engine* should contain both forward- and backward chaining search methods, and provide explanations of the results. An important feature is the possibility to consult external databases (GIS, etc). Finally, an AI shell should also provide the possibility of deriving rules from a set of examples (inductive knowledge).

The choice of a shell is of paramount importance for the success of an ES project: different problems require different reasoning processes (goal-driven search, data-driven search etc.) If the shell does not support the appropriate reasoning procedures it can lead to very cumbersome implementation procedures, and in the end even to wrong or inappropriate results.

It is obviously impossible here to list and discuss the merits of all of the AI shells today commercially available: therefore, we have included a list with commercial references in Appendix B.

5.6 - Example: conceptual construction of a Thermal-Plant-oriented ES

This Example describes one of the basic problems encountered in the development of an Expert System: *the formal implementation of the model rules*. Some of these rules (See Sects. 3.12 and 3.16 above) represent portions of the KB of the specific task(s) to be modelled, while others are expressions of a "general" or "specific" solution method. In the case discussed here, the objective was that *of explicitly formalising a set of definitions and of operational principles which could lead to the automatic synthesis of a feasible process starting from the specification of its product(s) and from the a priori imposition of some engineering constraints*. There is no unique way to do this: Section 2.1 above prescribes the general operational steps which are required to reach a solution, but no "universal" method has yet been found to formulate models of physical processes. The approach taken here is that of a *game*: we have tried to define some "rules-of-the-game" of modelling, starting with initial formulations which were both simple and "naive". These sets of rules have subsequently been sharpened and deepened by systematic use of "counterexamples", i.e., by that mathematical technique which goes under the name of *reductio ad absurdum*. The general procedure can be described as follows: first, an initial "common sense" definition of a rule is formulated, then it is implemented *mechanically and rigidly* to all types of practical engineering situations which can be imagined, trying to discover some absurd, non-physical or improper consequences of its application. The original rule is then iteratively amended, to avoid its "impossible" consequences and to make it more "robust" to slight perturbations in the engineering scenario with reference to which it was employed. This method may seem intrinsically slow, but it turns out that it usually converges rather quickly to feasible, physically sound, and logically robust rules with no implementation problems. It has to be remarked that at least *two* types of cultural backgrounds are needed in a team trying to follow this route: the physical/engineering know-how must be merged with its mathematical/philosophical counterpart. It is actually more convenient that the competence be totally separated, i.e., that the "mathematicians" do not know anything (or very little) about the physical processes to be synthesised. The problem is not that of understanding the principles of Engineering Material Science, Thermodynamics and Chemistry according to which a process is designed: they constitute the *domain knowledge* which can be easily captured in a set of possibly complex "Engineering Rules". The real problem is that *of capturing the formal principles which constitute the "Method" (or the "Art") of Engineering Design*. This is why the adoption of a "Design Game" strategy is convenient: because, since the rules of the game also define its outcome for a given set of premises, it is possible to adjust these rules to obtain the proper results from every other game one can think of playing under different premises.

The first analogy which we employed is that between a process and a mathematical proof. A proof is a finite sequel of propositions **p** such that either a **p** belongs to a set of fundamental axioms, or it can be deduced from other **p**'s in the proof via some strictly defined deduction rules. The process can be thought of as a finite sequel of fluxes **f** such that either a **f** belongs to a pre-defined "environment" or it can be produced from other **f**'s via a certain set of components each one of which performs a certain well-defined function ("elementary process").

Such an analogy -which we will accept without further scrutiny- bears two intrinsic weaknesses: first, a mathematical proof is *non-dissipative*, in the sense that its axioms and propositions can be re-used any finite number of times they are needed in the course of the application of the proof itself. Every real process implemented in any component *dissipates* some qualitative or quantitative property of the fluxes it "uses": therefore, fluxes are reusable only within limits. Secondly, while a proof has a *linear* structure, a process tends to display *loops* and *feedbacks*.

Even this basic analogy then must be used with care. But let us set excessive care aside by now, and try to develop the analogy further. Consider the following question: "When is a mathematical proof applicable?". The obvious answer is: "In all cases in which its premises are true". The analogy then would suggest that a component ought to be inserted in a process any time that the fluxes which it can work on are available: this is clearly an absurd result, in the sense that it does not lead to any "reasonable" process or plant. This side of the analogy cannot be applied: this rule does not belong to the game.

But, another answer to the question about the applicability of a mathematical proof is "A proof is applied:

a) if its premises are true
AND
b) if it brings about a reduction of complexity of the problem it is being applied to"

The analogy with an engineering process is more encouraging now: "A component is inserted in a process when:

a) the fluxes it works on are available
AND
b) it brings about a reduction of complexity in the process, i.e., it represents a step in the direction of the final product **F**".

This is a fundamental step in the "game strategy" we are trying to develop. In constructing a process, *we have to start from the final product and proceed backwards toward the "basic" fluxes of constituent materials*. The most "basic" materials are of course those contained in what we call "the environment", so one of the general rules of our game will be:

Rule #1: The process is to be thought of as a path connecting the final product **F** with the fluxes $\mathbf{f}_{env}$ available from the environment.

Now, the analogy can be carried farther. A mathematical proof can be represented by a graph (like a decision tree) in which the nodes are the successive "states" of the propositions which constitute the proof material, and the branches are the mathematical inferential symbols which connect these states ("implies", "negates", etc.). Similarly, a process can be thought of as a graph whose nodes are the components and whose branches are the fluxes. So, the second general rule of our game will be:

Rule #2: Process synthesis is *topologically* similar to the construction of a decision tree having its roots in the environment and whose "fruit" is the final product.

This constitutes another very important step in the search for a solution, but it has an important drawback: if the "fluxes/components tree" is indeed a suitable representation for a process, in the sense that it can *in principle* be used to proceed backwards from the final product to the environment, it can easily happen in practice that the extensive or/and the intensive parameters of some of the fluxes required by one component do not match the corresponding parameters of the "available" fluxes (at that level of the tree). Common engineering sense tells us that this is indeed the reason for inserting pumps, compressors, expanders, throttling valves and heat exchangers in real processes. So, Rule number 3 must be introduced:

Rule #3: A generic flux is specified (i.e., identified) both by its qualitative and quantitative parameters.

This rule assures us that a mismatch will be immediately detected. A first corrective action in such a case would be that of defining some "auxiliary" components (pumps, compressors, expanders, throttling valves and heat exchangers) and of inserting them on the skeleton of a process constructed by some "rules of inference" which lead from the product to the environment. But this is not what happens in reality: in the *real engineering game*, we are not only prepared in advance to insert, say, a feedwater pump upstream of a boiler to match the pressure level of the feedwater with that required by the boiler, but also to insert an air-to-water heat exchanger (intercooler) between the first and the second bodies of a large compressor, to improve the process even if the sensible heat of the water stream is subsequently wasted into the environment. The problem can be solved by recurring again to the "complexity" principle stated above: like a mathematical proof is applied whenever it reduces the complexity of the problem, so an auxiliary component should be inserted whenever possible, because its application introduces compatibility where there was none, and simplifies the process by avoiding the construction of more complicated alternatives to circumvent the mismatch. So, a fourth general rule of our game will be:

Rule #4: Whenever a mismatch is detected in the process parameters, auxiliary components must be introduced to eliminate it.

This rule is practically meaningless unless used in conjunction with yet another general rule:

Rule #5: The fluxes required as inputs by the auxiliary components must be sought first among those available from the environment and then among those made available by the process at its present stage of definition. Only if both of these searches are unsuccessful should new required external inputs be specified.

If we consider the set of game rules assembled so far, we realise that the entire "game" of process synthesis has been reduced to a top-down inference procedure, which starts from the final product (the final effect, the symptom) and traces its way back to the initial constituent materials (the first cause, the disease). The "local" sub-rules which drive the choice of the single components, of the type and structure of the connections, etc., are the usual engineering rules and can be easily implemented in "local" procedures.
The problem is solved, at least in principle. Or, is it? Again, the game approach offered some further insight. This type of procedure can be applied without excessive difficulty to the synthesis of energy conversion processes, as long as they do not require extensive use of heat exchangers chains (also called Heat Exchangers Networks, "HEN"). Repeated failures in designing a heat exchanger chain led to the discovery that very often, the top-down strategy alone cannot handle these cases. When several streams, some "hot" and some "cold", must exchange heat among them in a thermodynamically acceptable fashion, a compromise must be struck between the top-down and the bottom-up approach. Suppose, in fact, that the first of these heat exchangers is introduced. It will fix one or more mismatches, and produce a "less hot" and a "less cold" outputs than they were before its insertion in the process. But, what if we now introduce a second heat exchanger? It, too, will reduce some mismatches and modify the temperature levels of some of the fluxes. Would it have been the same if we had inserted the second heat exchanger first? Almost surely, not so: a heat exchanger reduces a mismatch in a different way depending on the stage of the layout at which its insertion takes place. But this is tantamount negating the top-down approach! It becomes necessary at this point to introduce yet another game rule:

Rule #6: Any time two or more heat exchangers are being inserted in a process diagram, even if the insertion of each additional unit is decided according to a top-down strategy, the procedure to be used is of a bottom-up type. Since every new heat exchanger is subjected to local condition which depend on the present stage of the process layout, a "local" iterative procedure must be introduced to fine tune the heat exchanger chain.

This rule appears once more to cure all problems. Experience has shown, however, that it is difficult to avoid "non-local" effects of the iterative procedure. Often, the interconnectivity of a process is so high, that small changes propagate throughout the entire structure very quickly (i.e., within few iterations). This is not *-per se-* a negative feature, and it is actually desirable that a process synthesis procedure be so sensitive: but the number of slightly different processes which are originated by this strategy grows exponentially with the number of heat exchangers, and can become so taxing for the computational resources to prevent the code from producing any output within reasonable times. Yet another fix is necessary. The (tentative!) solution emerged again from an analysis of the thinking patterns of designers. It came to our attention that, when choosing a component, no exact sizing is done by the Process Engineer: rather, an *approximated* or *estimated* value for the component "size" is guessed, and some corresponding values for the relevant fluxes connected to that component are derived. It is only at a later stage in the process design that balance equations are used to completely specify all components. The difference between an "expert" engineer and a novice consists mostly in the fact that the "estimates" of the expert are much closer to reality than those of the novice. A similar procedure can be introduced in our game strategy: each component can be labelled by an "approximate" indication of its size, and the entire layout can be devised enforcing an *approximate match* between all components, including heat exchangers. The number of acceptable solutions will increase substantially, but the advantage of reduced iterations will probably more than offset this effect. And, once a certain number of complete process structures has been created, it is not difficult to run a mass- and energy balance code to exactly size each component. This brings us to the formulation of another game rule:

Rule #7: When introducing a new component in a process, a *fuzzy label* must be attached to it, which contains the estimated operational range for both its extensive and intensive parameters. The immediate "successors" of this component (in the top-down direction which the procedure commonly follows) must have operational ranges compatible with those of their parent component. This rule applies to heat exchangers as well.

The seven game rules defined so far seem to constitute a successful game strategy. Further verification is pending (see Examples 6.6 and 6.7). We are still playing this game.

CHAPTER 6

EXAMPLES OF APPLICATIONS OF EXPERT SYSTEMS TO THE DESIGN AND CONTROL OF ENERGY CONVERSION SYSTEMS AND COMPONENTS

6.0 - FOREWORD

The nine examples included in this Chapter are representative of the possible uses of Expert Systems in the field of Thermal Systems Engineering. The first three relate to applications made on existing plants, and can be thus characterised as dealing with *plant operation problems*; the next four (4 through 7) are direct attempts to solving *design problems*: examples 4 and 6 deal with the choice of a component and of a system respectively, example 5 is a direct application to quantitative component design, and example 7 is an application to Process Synthesis. The last two examples (8 and 9) are explicitly related to the concept of "Optimal Design", and are -in the view of the authors- worth careful consideration: they represent attempts to generate a feasible process- or plant configuration with no previous qualitative or quantitative knowledge of "what the solution should look like", basing the decision process on qualitative general "common sense engineering rules" only.

These examples should serve as a guide to the reader: they do not represent the *only* solution to the problem they tackle, nor claim to represent an *optimal solution* (from the point of view of the coding or of the logical structure); but they are so presented to give an exact and immediate idea of *what* can be done to solve the given engineering problem, and of *how* it can be done.

EXAMPLE 6.1 - DETECTION AND CORRECTION OF COMPRESSOR STALL AND SURGE

6.1.1 - THE PHYSICAL PROBLEM

Compressor stall and surge are two different but interrelated malfunctions which may arise during otherwise normal operation of a compressor, and which, besides seriously affecting its performance, can have devastating structural effects on the blades, bearings and, in extreme cases, on the shafts of the turbomachines. Therefore, their early detection and correction is an important task of the Plant Engineer[1].

The physics of both phenomena is complex, and its complete description is outside of the purpose of this example: interested readers are referred to [28]. Here, we will only describe *the general features of stall and surge*, and touch upon their fluid dynamic analysis only for gathering enough information to plan for proper corrective action. With this understanding, the problem can then be posed as follows: based on the available phenomenological and theoretical knowledge, devise an automatic "intelligent" system capable of detecting incipient stall and surge, of taking corrective action, and of alerting the plant operator.

6.1.2 - SOME THEORETICAL CONSIDERATIONS

This is an example of operator-oriented ES. It should perform a corrective action in response to real events in a real operational environment, by "substituting", in real or virtual terms, the human operators presently supervising the process. A general description of the tasks performed by a human operator is illustrated in Fig. 6.1, and consists of:

- collecting and evaluating the information presented by the process/human interface (the sensors' displays);
- performing low-level activities ("operational level interventions") by direct application of the operator's manual;
- recognising a first meta-level of intervention, called the *tactical level*, in which all decisions are taken on the basis of some well-defined rules partly written in the operator's manual and partly acquired by experience. Possibly, these rules may require a substantial amount of trial-and-error techniques;
- recognising a second meta-level of intervention, called the *strategic level*, in which all decisions are taken on the basis of some possibly fuzzy rules, the

[1] There is also a design side of this story, because both stall and surge can be prevented (at least to a certain extent) by proper design of the machine: we will not consider this issue here.

most of which are not contained in the operator's manual, inferred by experience and requiring a limited use of trial-and-error techniques;

- being able to handle with the utmost celerity abnormal operational conditions (which constitute "exceptions" to the rules)

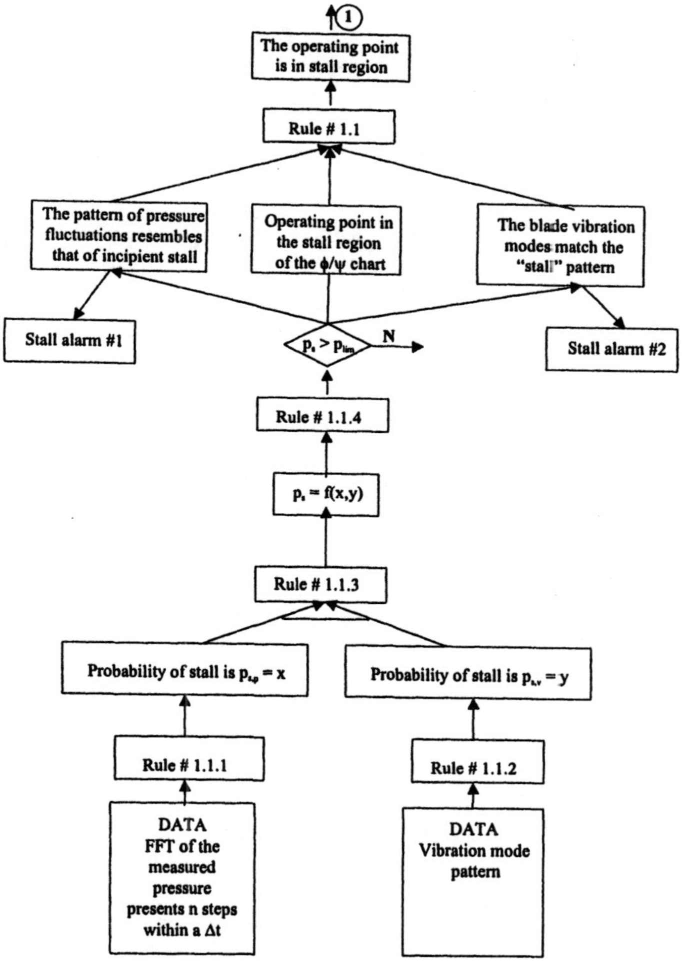

Figure 6-1. Backward chaining inference mechanism

Often it is not a matter of "response speed", but of accuracy and pertinence of the response. Perhaps the most striking quality of the operator's intervention is that it constitutes the control equivalent of a reverse direct problem in design. In fact, while operation of a plant is usually established by top-down procedures, because the control strategy has been devised by mimicking the mental model of the Design Engineer, and it is essentially based on top-level knowledge applied to the management of low-level data, the operator's intervention is performed in response to the prompt "do what you have to do to achieve (or to restore) goals A and B on the basis of your knowledge of both the logical and physical structure of the process and of its functional response to variations of the boundary conditions or of operating parameters". The ES should of course perform according to a logical scheme somehow reproducing that consciously or unconsciously employed by human operators: the steps are made explicit in Fig. 6.2.

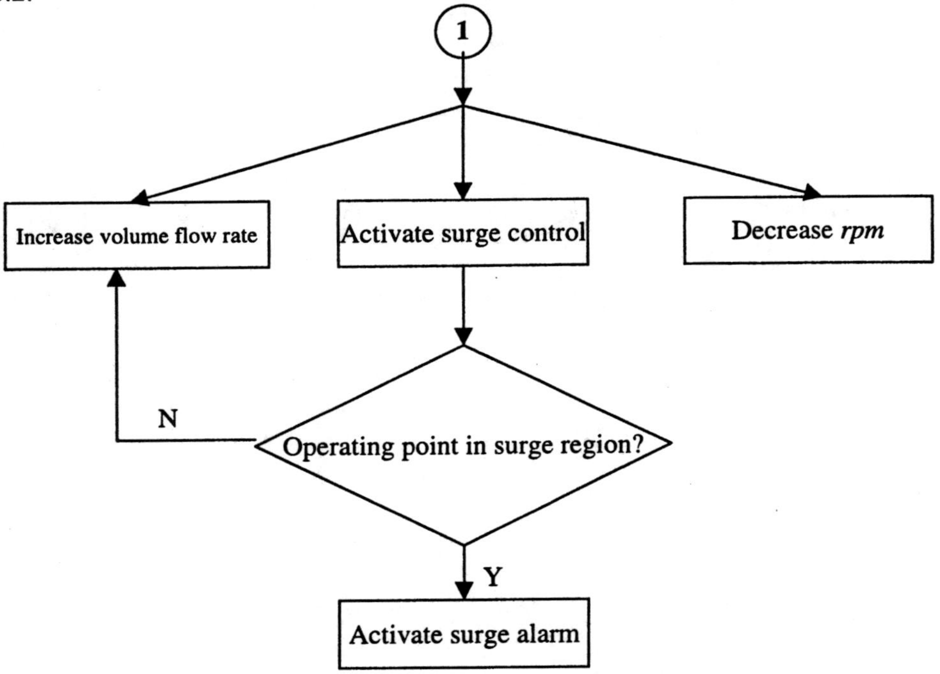

Figure 6-2. Expert System intervention tasks

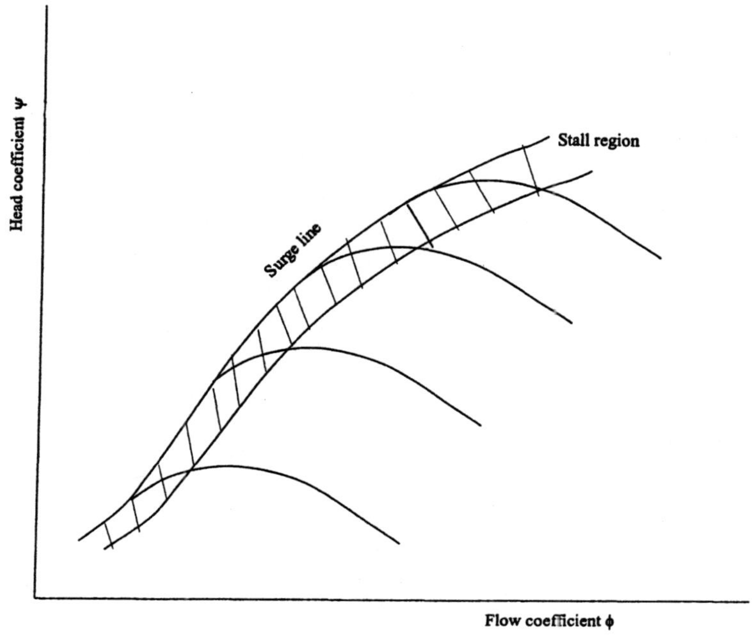

Figure 6-3. ϕ/ψ chart

6.1.3 - SOLUTION

Let us assume that the necessary engineering understanding of the phenomena is available from a *knowledge acquisition and analysis* activity which has resulted in the following Knowledge Base[2]:

- a "surge curve" is identified experimentally and can be expressed by a functional relationship between pressure rise, inlet flow and rpm (see Fig.6.3);
- both stall and surge can be identified as "regions" in the space defined by ϕ/ψ charts (see Fig. 6.3);
- sure symptoms of the onset of stall are well-defined patterns of pressure fluctuations, like those depicted in Fig. 6.4;

[2] The list of "symptoms" is by no means to be interpreted as exhaustive, and is only employed here as an example

- additional symptoms of the onset of stall are certain fluctuations in the mass flow rate, also depicted in Fig. 6.4;

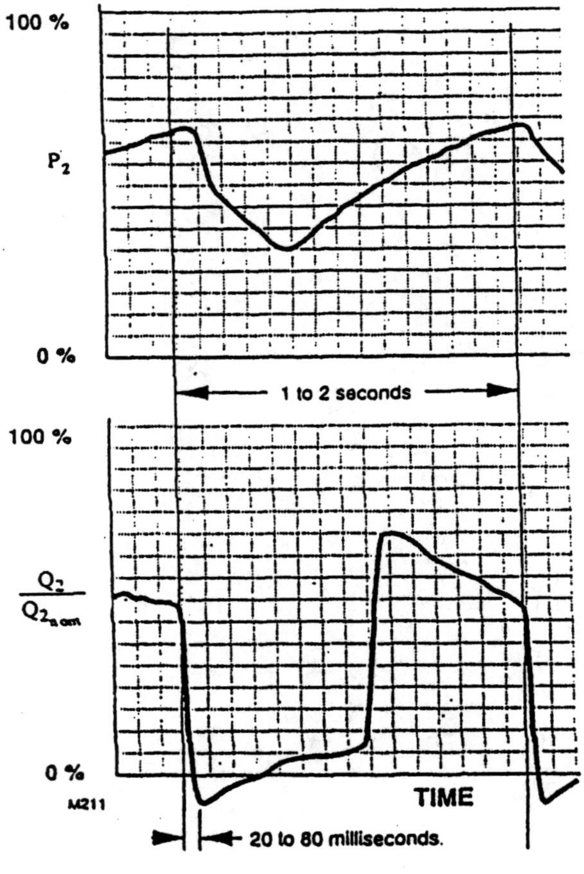

Figure 6-4. Typical stall-induced pressure and flow-rate fluctuations

- tip- or rotating stall result in certain well-defined blade vibration modes;
- surge (both mild and abrupt) is *always* preceded by some form of stall;

- incipient surge is characterised by typical peaks in pressure fluctuations, whose shape and frequency depends on the exact features of the phenomenon (rotor stall, diffuser stall);
- surge also results in typical local temperature rises due to extensive flow reversal and to the resulting reduction in compressor efficiency;
- macroscopic vibrations and the corresponding noise usually result from the *persistence* of surging conditions;
- the initial "high-pitch" noise which is often taken as a surge-warning signal is due to the turbulent bursting of rotating stall cells, and is indeed a sure symptom for the onset of abrupt surge.

The means for correcting the operating mode of the machine and for bringing it back to normal operation are essentially four:

1. *bleed*: the mass flow rate is decreased by opening up some bleeding (or recirculation) slots between the stages (in axial compressors) or along the channel (in centrifugal compressors). The purpose is to bring the operating point back to the right of the stall or surge line (see Fig. 6.3);
2. *variable inlet guide vanes*: absolute axial flow velocity rate is decreased at the inlet by inducing preswirl in the incoming flow. This modifies the angle of attack and prevents or strongly reduces stall onset;
3. *variable diffuser blades*: a change in the relative angle of diffuser blades with respect to the absolute velocity reduces diffuser stall and is also very effective in eliminating surge, because it modifies the dynamic "impedance" of the downstream plenum (as seen by the compressor);
4. *variable outlet throttling*: a throttling device downstream of the diffuser helps reduce surge, by modifying the downstream impedance.

This relatively simple set of knowledge data can be used to devise a control strategy based on an inductive process which tries to go back to the *causes* starting from the detected *effects*. The rules governing the inference engine will be of the general type:

IF p THEN q

where **p** is the *activation condition* (it can consist of more than one rule), and **q** is a series of stall- and surge preventing or correcting activities.

As an example, consider the following two macro-rules:

rule \#1): IF operating point is in stall region
THEN increase inlet volume flow rate
OR introduce preswirl
OR decrease *rpm*
AND activate surge control devices

rule \#2): IF operating point is in surge region
THEN increase inlet volume flow rate
OR introduce preswirl
AND activate surge alarm

Notice that both the premise and the conclusion of each of the two rules is constituted by *composite* instruction: in fact, the premise of rule \#1 can be expanded into:

rule #1.1): IF operating point is in the "stall" region of the ϕ/ψ chart
OR the pattern of pressure fluctuations resembles that of incipient stall
OR the blade vibration modes match the "stall" pattern
THEN operating point is in stall region

In turn, the two "OR" in rule \#1.1 can be combined and expanded to introduce some probability consideration:

rule #1.1.1): IF FFT of the measured pressure presents *n* steps within a certain Δt
THEN probability of stall is $p_{s,p} = x$

rule #1.1.2): IF vibration mode pattern is of a type compatible with stall
THEN probability of stall is $p_{s,v} = y$

rule #1.1.3): IF 1.1.1 **AND** 1.1.2 are true
THEN probability of stall is $p_s = f(x,y)$

rule #1.1.4): IF $p_s > p_{s,limit}$
THEN operating point is in stall region

6.2 - Emission Monitoring and Control of a Fossil-Fueled Powerplant

6.2.1 - The Physical Problem

Consider the following plant control problem, and construct a decision tree for it. For a fossil-fuelled powerplant, it is necessary to adjust the power output *in operation*, according to the instantaneous values of four chemicals emitted by the plant, namely O_2, CO, NO_x and SO_2. The only controlling variables are (see Fig. 6.5):

- the air mass flow rate, m_a;
- the fuel mass flow rate, m_f;
- the mass flow rate of the water injected into the combustor, m_w;
- the mass flow rate of the water sprayed into the semi-dry hot gas scrubber, m_s;

The pollutants' concentrations in the stack gases, as well as the residual oxygen, are measured continuously, and 5-minutes averages are available. Assume that the response time of the plant is 1/2 hour (i.e., a step variation of one of the input variables at t_o will result -after a 1/2 hour transient- in a steady-state variation of the output conditions).

The regulatory limits are:

$O_2 \geq 6$ %
$CO \leq 100$ mg/mc
$NO_x \leq 300$ mg/mc
$SO_2 \leq 50$ mg/mc

6.2.2 - SOME THEORETICAL CONSIDERATIONS

This is an operator's task again, but it is much more complex than the one described in Example 6.1. Here, the knowledge necessary to correctly perform the control actions is almost entirely at the second level, the tactical one (see Fig. 6.6), and some is even at level three, because it involves strategic decisions about the plant's outputs. Otherwise, the same considerations apply here as to Example 1 above.

6.2.3 - SOLUTION

Let us assume that some knowledge on the *operator's interventions* necessary to control the emissions is available from previous experience, and that it can be represented in the form of *flow-charts*, as in Figs. 6.6a through 6.6d.

At first glance, it seems possible to translate the flow-charts into a set of *conditional* rules of the form:

IF p THEN q

where **p** is the *activation condition* (it can consist of more than one rule), and **q** is a well-defined series of operator's interventions.

For instance, one could formulate the following two rules to control the NO_x emission levels:

rule #1): IF $NO_x > NO_{x,max}$ **AND** $T \neq T_{opt}$
THEN optimize T acting on m_a and m_f

rule #2): IF $NO_x > NO_{x,max}$ **AND** $T \approx T_{opt}$

THEN increase m_w

Actually, a third rule ought to be inserted here for safety's sake:

rule #3): IF $NO_x > NO_{x,max}$ **AND** m_w cannot be increased
THEN activate alarm and prompt operator

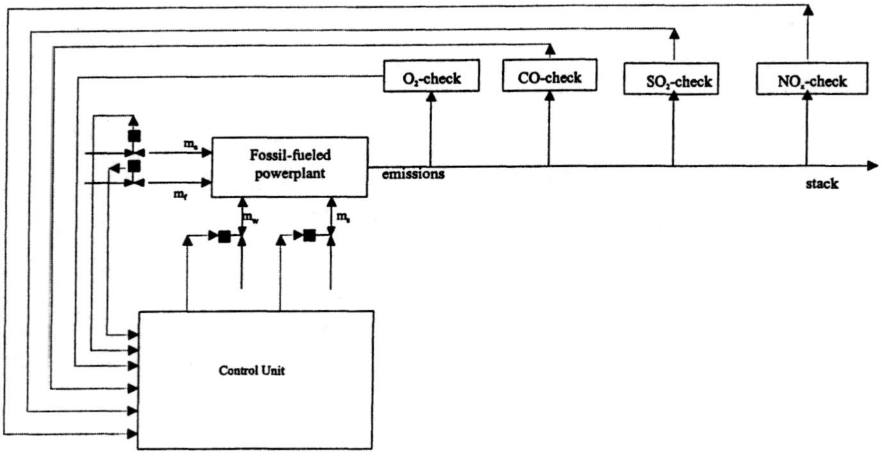

Figure 6-5. Control scheme for a fossil-fueled powerplant

This simple set of rules needs to be "steered" by another set of *priority rules*, which controls their activating sequence. In this case, we can safely assume that rule #1 is prioritary with respect to both rule #2 and #3, while rule #2 is prioritary with respect to rule #3.

To complete the control sequence, we need only to introduce yet another rule, the *repetition rule*, which allows the control sequence to repeatedly enforce either one of the rules, if its premises (the predicate of the "IF...") continue to apply.

From the remaining flow charts, similar rules can be derived to control the emission level of the remaining pollutants:

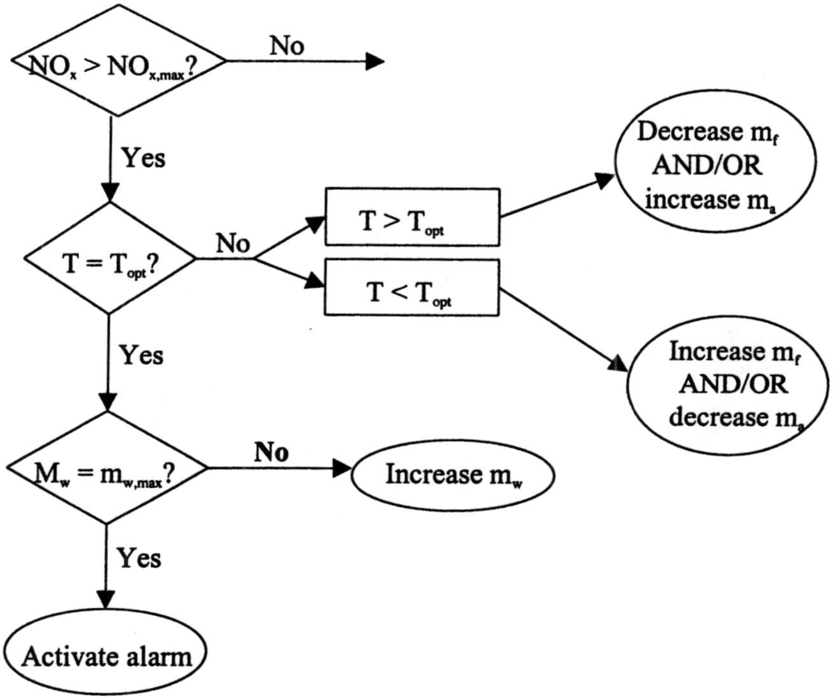

Figure 6-6a. Flow-chart for NO_x control

Figure 6-6b. Flow-chart for SO_2 control

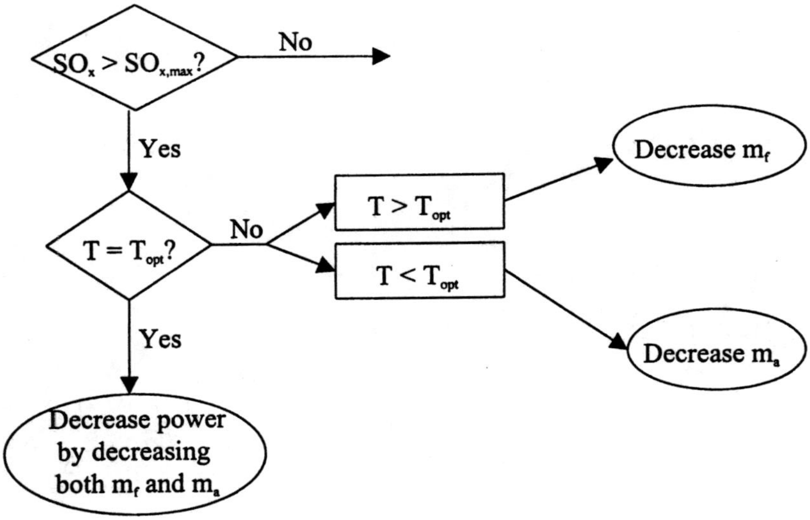

Figure 6-6c. Flow-chart for O_2 control

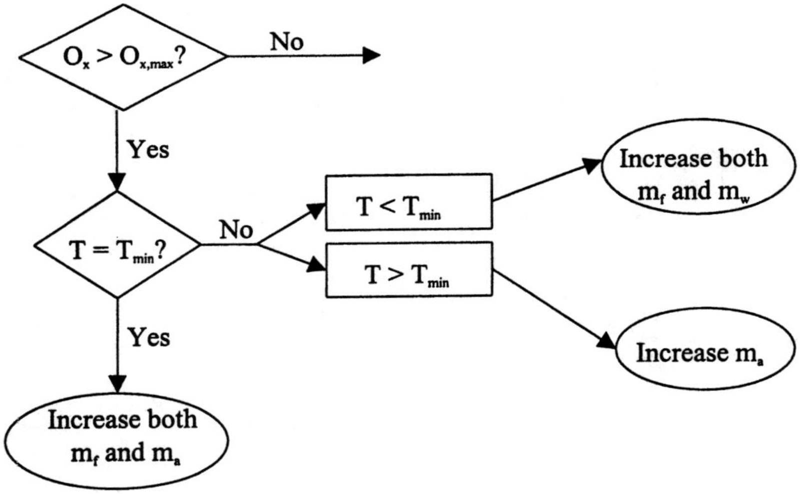

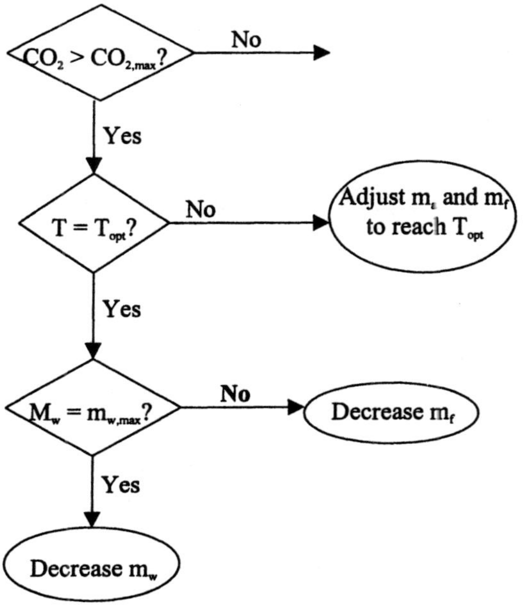

Figure 6-6d. Flow-chart for CO_2 control

rule #4): IF $SO_2 > SO_{2,max}$ **AND** $T \neq T_{opt}$
THEN optimise T acting on m_w and m_f

rule #5): IF $NO_x > NO_{x,max}$ **AND** $T \approx T_{opt}$
THEN increase bypass ratio

rule #6): IF $NO_x > NO_{x,max}$ **AND** bypass ratio cannot be increased
THEN activate alarm and prompt operator

rule #7): IF $O_2 < O_{2,min}$ **AND** $T < T_{min}$
THEN increase m_w and m_f

rule #8): IF $O_2 < O_{2,min}$ **AND** $T > T_{min}$
THEN increase m_a

rule #9): IF $O_2 < O_{2,min}$ **AND** (m_w **OR** m_a) cannot be increased
THEN activate alarm and prompt operator

rule #10): IF $CO > CO_{max}$ **AND** $T \geq T_{max}$
THEN reduce power

rule #11): IF $CO > CO_{max}$ AND $T < T_{max}$ **AND** $T \approx T_{opt}$
THEN increase m_w **AND** maintain constant power

rule #12): IF $CO > CO_{max}$ **AND** $T < T_{max}$ **AND** $T \neq T_{opt}$
THEN adjust m_a and m_f to reach T_{opt}

rule #13): IF $CO > CO_{max}$ **AND** (m_w **AND** m_a **AND** m_f) cannot be modified
AND power cannot be reduced
THEN activate alarm and prompt operator

The 13 rules, properly augmented with their respective priority and repetition rules, would seem to solve the problem. A more accurate analysis would show (the reader can easily check this by simple inspection) that the control sequence can enter in some *loop* and become ineffective under certain operating conditions. For example, if rule #8 is enforced, the increase of m_a which it causes could conflict with rule #12, and the system would go back and forth between the two rules forever, because it does not have any criterion as to which one of the pollutants to reduce first. So, one possible solution could be that of introducing an additional rule which would assign this priority.

But deciding which pollutant is the "less dangerous" is definitely not a winning strategy if the name of the game is "abide *exactly* to *all* environmental limitations". The problem is, the different control sequences which are effective if applied separately *must be overseen by a macro-control sequence* which keeps track of the *real* scope of the system, which is that of *maintaining the emission level below the mandatory standards, while maximising the power output from the plant.* A hint on how to implement such a system is obtained by an analysis of the behaviour of an expert human operator: s/he would *execute the single control actions sequentially, trying to avoid those which conflict with the ones already activated.* This behaviour can be mimicked by an automatic control system consisting of the following rules:

rule #1): IF $SO_2 > SO_{2,max}$
THEN decrease SO_2

rule #2): IF $NO_x > NO_{x,max}$
THEN decrease NO_x

rule #3): IF $O_2 < O_{2,min}$
THEN increase O_2

rule #4): IF $CO > CO_{max}$
THEN decrease CO

rule #5): IF CO cannot be decreased **AND** SO_2 cannot be decreased
AND NO_x cannot be decreased **AND** O_2 cannot be increased
THEN reduce power **AND** activate alarm **AND** prompt operator

It is clear that the predicates of the various "THEN" ("decrease NO_x", etc.) will be complex rules, i.e., they will consist of several logical instructions which, taken together, articulate their goal. For example, " decrease CO " will consist of the following sub-rules[3]:

IF $CO > CO_{max}$ **AND** $T \geq T_{max}$ **AND** "reduced P = not true"
THEN reduce power **AND** attach flag "reduced" to variable P

IF $CO > CO_{max}$ **AND** $T < T_{max}$ **AND** $T \approx Topt$
AND "reduced m_w = not true"
THEN increase m_w **AND** attach flag "reduced" to variable "m_w" **AND** maintain constant P **AND** attach flag "constant" to variable P

IF $CO > CO_{max}$ **AND** $T < T_{max}$ **AND** $T \neq T_{opt}$
AND "(reduced m_a **AND** reduced m_f) = not true"
THEN adjust m_a **AND** m_f to reach T_{opt}

IF $CO > CO_{max}$ **AND** (m_w **AND** m_a **AND** m_f) cannot be modified **AND** power cannot be reduced
THEN activate alarm and prompt operator

With this structure, the control system will take sequentially[4] each control action, and will resolve conflicts by elimination (because, for instance, if m_a has been decreased to reduce NO_x, it will not be increased again to reach the minimum required O_2). Of course, the emergency situation in which the control sequence fails are reported to the operator.

Notice that, if the relationship between the emission levels and the control variables is amenable to an explicit formalization, then the knowledge about the system's behaviour can be expressed in a very compact and effective fashion, by means of the so-called *influence matrix*. Suppose that the emission levels are expressed, in the space spanned by m_a, m_f, m_w, P and T, by explicit formulas:

$$NO_x = f(m_a, m_f, m_w, P, T)$$
$$SO_2 = g(m_a, m_f, m_w, P, T)$$
$$CO = h(m_a, m_f, m_w, P, T)$$
$$O_2 = i(m_a, m_f, m_w, P, T)$$

Suppose further that the derivative of the emission level of each pollutant can be calculated and the following relations exist:

---

[3] "P" denotes mechanical power in the following

[4] The sequence in which CO, O_2, NO_x, SO_2 are controlled could influence the effectiveness of the control chain, but we will not delve on this aspect of the problem here

$$\delta f / \delta T > 0 \text{ for }, T < T_{opt}$$
$$< 0 \text{ for }, T > T_{opt}$$

$$\delta f / \delta m_a > 0 \text{ for } m_a < m_{a,design}$$
$$< 0 \text{ for } m_a > m_{a,design}$$

$$\delta f / \delta m_f < 0 \text{ for } m_f < m_{f,design}$$
$$> 0 \text{ for } m_f > m_{f,design}$$

$$\delta f / \delta m_w < 0 \text{ for } \forall \ m_w$$

$$\delta f / \delta P > 0 \text{ for } P > P_{design}$$
$$< 0 \text{ for } P < P_{design}$$

Then, the behaviour of NO_x for what its control is concerned can be expressed by a "table of values" of its partial derivatives:

	$\delta f / \delta T$	$\delta f / \delta P$	$\delta f / \delta m_a$	$\delta f / \delta m_f$	$\delta f / \delta m_w$
$T > T_{opt}$	-	0	0	0	0
$T < T_{opt}$	+	0	0	0	0
$P > P_{des}$	0	+	0	0	0
$P < P_{des}$	0	-	0	0	0
$m_a > m_{a,des}$	0	0	+	0	0
$m_a < m_{a,des}$	0	0	-	0	0
$m_f > m_{f,des}$	0	0	0	+	0
$m_f < m_{f,des}$	0	0	0	-	0
$m_w > m_{w,des}$	0	0	0	0	-
$m_w < m_{w,des}$	0	0	0	0	+

This table is called the *influence matrix* of the problem for the variable NO_x, and gives a clear indication about what control action to take in different regions of the control space. Similar tables can be constructed or each relevant variable.

6.3 - AUTOMATIC OPERATION OF A FUEL-STORAGE FACILITY

6.3.1 - THE PHYSICAL PROBLEM

The fuel-storage facility shown in Fig. 6.7 is presently manually controlled by an operator who can choose among different operational modes:

A) *storage pumping liquid fuel to plant*

A.1- unloading tank no. 1
A.2- unloading tank no. 2

B) *storage pumping gaseous fuel to plant*

B.1- unloading tank no. 3
B.2- unloading tank no. 4
B.3- unloading tank no. 5
B.4- unloading tank no. 6

C) *storage being filled with liquid fuel from ship at dock*

C.1- filling tank no. 1

C.2- filling tank no. 2

D) *storage being filled with gaseous fuel from ship at dock*

D.1- filling tank no. 3
D.2- filling tank no. 4
D.3- filling tank no. 5
D.4- filling tank no. 6

The operator can select the mode, and execute the proper control sequences, by acting, via a series of PLC-assisted circuits, on:

- the loading pumps, P_1 and P_2;
- the loading compressor, C_1 and C_2;
- the discharge pumps P_3 and P_4;
- several control and pressure valves.

It is required that an ES be designed to replace the human operators.

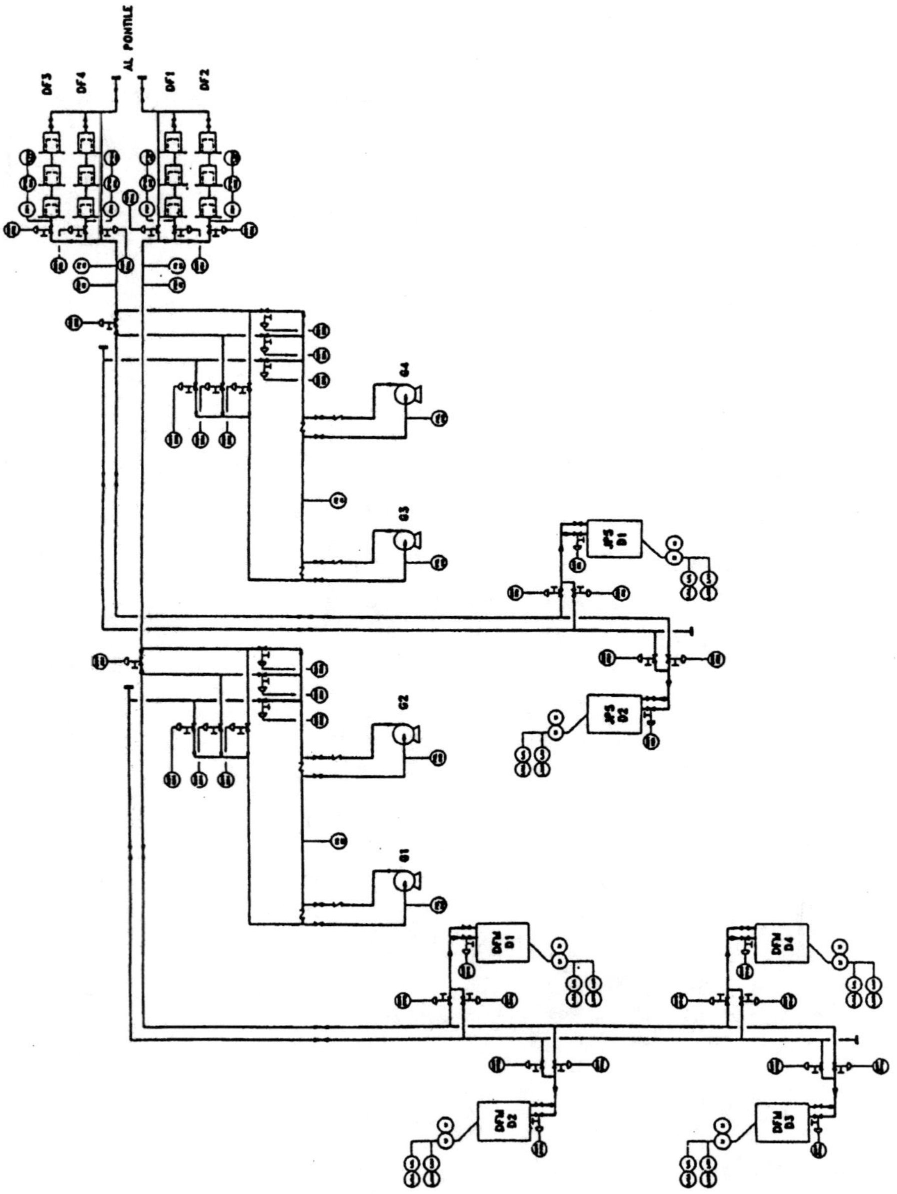

Figure 6-7. Process and Installation diagram for a fuel storage facility

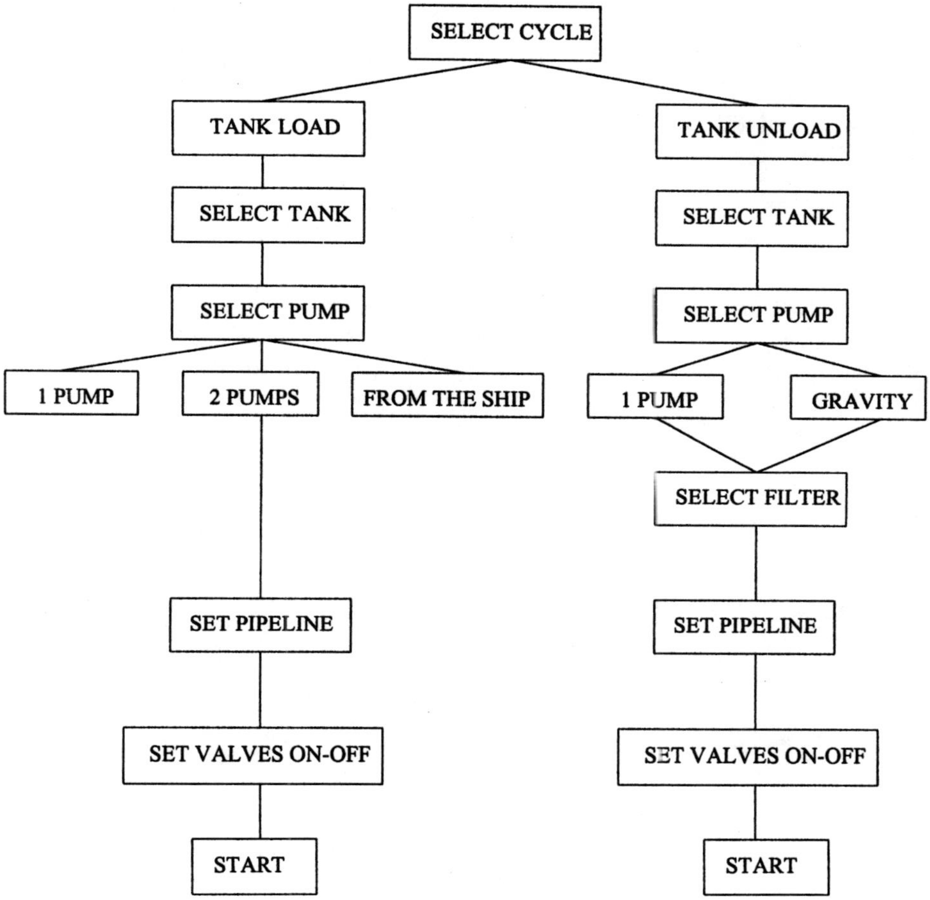

Figure 6-8a. Decision tree for fuel tanks loading/unloading operations

6.3.2 - SOME THEORETICAL CONSIDERATIONS

This is again a control problem, but it turns out to be simpler than the two previous ones. Its solution can be easily found by a careful analysis of the available knowledge (including the rules about possible connections and fuels incompatibilities) and the compilation of a Truth Table.

6.3.3 - SOLUTION

Let us assume that some knowledge on the operator's interventions necessary to control the loading/unloading operations is available from a previous knowledge analysis, and that it can be represented in the form of a flow-chart, as in Fig. 6.8.

It is possible to translate the flow-chart into a set of conditional rules of the form:

IF p THEN q

where **p** is the *activation condition* (it can consist of more than one rule), and **q** is a well-defined series of operator's interventions. For the purpose of the present analysis, both **p** and **q** can be *composite rules*, i.e., they can contain a certain number of sub-rules. In general, the control task will thus consist of the following instructions:

rule #1): **IF** operating mode = Filling **AND** constraints = C_i
THEN use control sequence F_i

rule \#2): **IF** operating mode = Unloading **AND** constraints = C_j
THEN use control sequence U_j

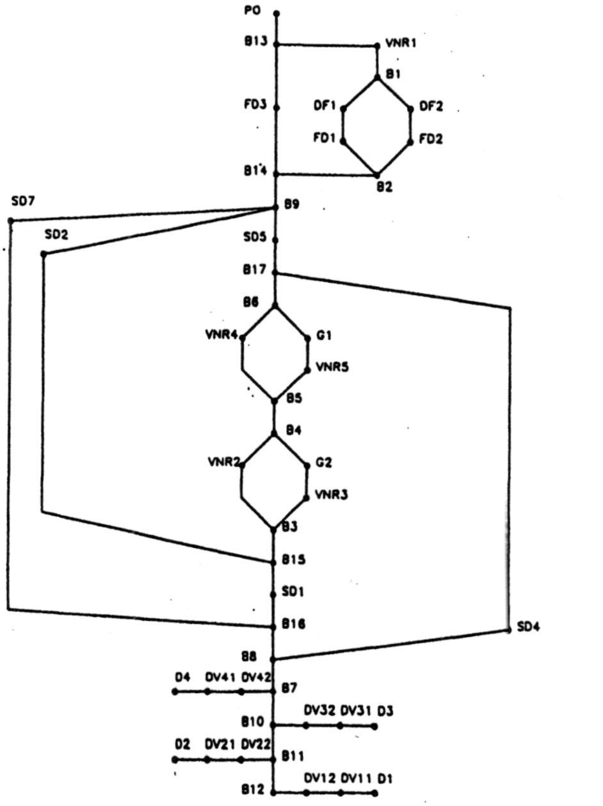

B = T-junctions
DF = filter
G = pump

D = Tank
FD = Motorized valve for filter unit
SD = Motorized valve for pipeline

DV = Tank motorized valve
PO = Dock
VNR = Check valve

Figure 6-8b. Graph representation for the tank loading/unloading operations

It is clear that the predicates of the various "THEN" ("open valve V_n", "control capacity of tank number m", etc.) will consist of several logical instructions which, taken together, articulate their goal. For example, "*fill tank number m with liquid fuel*" will consist of the following sub-rules (unit numbers are arbitrary):

IF (quantity of fuel to be loaded) ≤ (available capacity of tank m)
THEN open valves 1,3,6 and 8 **AND** close valves 2,4,5 and 7
AND activate pumps 1 and 2

With this structure, the control system will take sequentially each control action, and will resolve conflicts by elimination (because, for instance, if valve 1 has been opened, it cannot be closed again during the same operation unless specifically requested by the active control sequence). Of course, any emergency situation in which the control sequence appears to fail are to be reported to the operator.

In most cases, the knowledge about the system's behaviour can be expressed in a very compact and effective fashion, by means of the so-called *influence matrix* (otherwise called *truth table*). Suppose that the control actions are only of the form "on/off": then, the sequel of operations can be expressed by a binary table of values of the status of the components of the system:

component	tank 3 filling	tank 4 filling	tank 5 filling	tank 6 filling
valve 1	on	on	off	off
valve 2	off	on	off	on
valve 3	on	off	on	off
valve 4	off	off	on	on
....	....	....	....	....
compressor 1	on	on	off	off
compressor 2	off	off	on	on
....	....	....	....	....

This table, together with all other tables applying to the remaining set of possible operating conditions (filling tanks 1 and 2, unloading the tanks), gives a clear indication about the control strategy.

6.4 - Choice of a Feedwater Pump

6.4.1 - The Physical Problem

It is required that an automatic procedure be devised for the choice and specification of a feedwater pump. Since the extra work is really negligible, it will be convenient to construct the procedure so that it be capable of handling any kind of pump geometry (axial, mixed, centrifugal), and of covering the entire range of possible operation (see [28] for details).

The following is a "task list" which can be constructed by consulting a pump design textbook (or by asking an expert!):

- read design data: Q, H
- read constraints (if applicable): NPSH, speed, service, etc.
- is rpm given?

if Y: compute n_s
if N: is NPSH given?
if Y ⇒ compute iteratively s and derive rpm_{max}
if N ⇒ (assume rpm, compute NPSH, check n_s),
- check if multi-staging is required
- check if multiple pumps in parallel are required
- compute d_s from Cordier line
- compute D_2
- choose $\phi_T = D_2 \; U_2$
- compute b_2
- choose $D_{le} \quad D_{le}$
- compute D_{1e} and D_{1i}
- verify torsion stress on shaft
- compute overall dimensions (approximate)
- check component library for nearest defined pump type
- check characteristics: head at Q_{design}, Q_{min} and Q_{max}
- produce a technical specification sheet

6.4.2 - SOME THEORETICAL CONSIDERATIONS

This is *selection problem*, which is one of the simplest forms of a design task, and certainly requires manipulation of both qualitative and quantitative knowledge. Since the technology of both Hydraulic and Mechanical Engineering is very mature for what concerns pumps, it is likely that many viable designs exist, and that the task could actually be redefined as "selecting one among *n* possible objects that possesses the attributes specified in the problem position and covers the required field of operation". The simplest solution strategy is that of *elimination*: we can define some characteristic indices that qualify each different type of pump, compute their numerical or logical values which are determined by the problem position, and exclude all those types that do not match these values. Notice that some of these indices are qualitative ("fluid is corrosive", "fluid is hot", etc.).

6.4.3 - SOLUTION

The flow chart of the engineering design choices is depicted in Fig. 6.9. We will try to reduce the tree to a sequel of rules of the type:

IF p THEN q

where **p** is the premise (which can consist of more than one rule), and **q** is a series of design activities.

In this example, the decision tree (which we will assume to have been drawn following some "expert" advice) shows that, in addition to conditional rules of the type shown above, there are some other steps which imply some quantitative decision, like "*choose* $NPSH_{min}$, *choose* D_1, *etc.*". The easiest way to account for these steps is to treat them as *functional calls* to some external numerical procedure, which will return the values of the properly addressed variables. Keeping this in mind, a set of rules which reproduces the decision tree of Fig. 6.9 is the following:

rule #1): **IF** rpm not given
THEN call CAVIT(NPSH,s,rpm,T) **AND** read resulting rpm value

rule #2): **IF** pump type not assigned
THEN call PTYPE(rpm,Q,H) **AND** read n_s and resulting pump type

rule #3): **IF** n_{stages} not assigned
THEN call PSTAGE(n_s,Q,H) **AND** read resulting n_{stages}

rule #4): **IF** $n_s(Q,rpm,H_{stage}) \neq n_s$,old
THEN call CAVIT **AND** call PTYPE **AND** call PSTAGE **AND** iterate

rule #5): **IF** dimensions are not constrained
THEN call PSHAPE(Q,H_s,n_s) **AND** read D_1,D_2, etc.

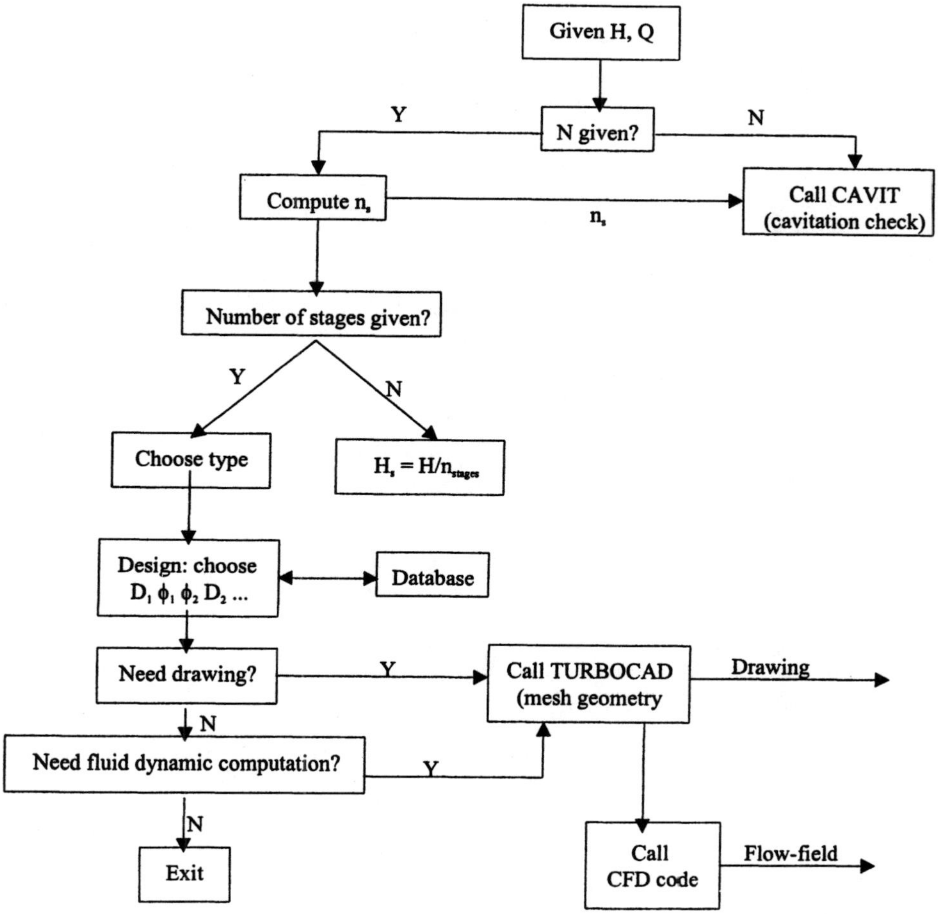

Figure 6-9. Decision tree for pump selection

The remaining steps can be executed conditionally at the user's prompt.

As an example, the PTYPE function could be expressed by the following "truth table" ($n_s = \frac{rpm}{H^{0.75}}$):

n_s	pump type
<0.2	multistage centrifugal
<2.5	centrifugal
<3.5	elico-centrifugal
<5.5	axial
>5.5	multiple axial in parallel

As already seen in Example 6.1 above, the use of functional maps for the pump greatly facilitate the execution of the procedure. "Maps" for the choice of the type of pump are readily available (see Fig. 6.10), as well as tables from which one can choose D_1,D_2,ϕ_1 etc. as functions of n_s. This is not a marginal point, nor it is by chance that such information has been codified for us in this form: these maps and tables *represent the combined knowledge of a multitude of experts which have solved the same problems before us*, and constitute in this case a conveniently structured, ready-to-use *Knowledge Base.*

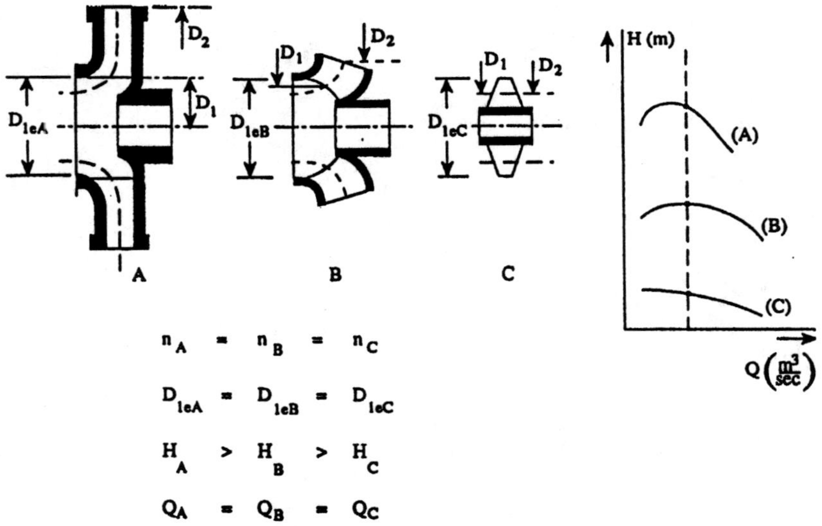

Figure 6-10. Schematical operational maps for different types of pumps

6.5 - CHOICE AND DESIGN OF A SHELL-AND-TUBE HEAT EXCHANGER

6.5.1 - THE PHYSICAL PROBLEM

The objective of this task is that of designing a Shell-and-Tube Heat Exchanger under the following requirements:

h) the two streams between which the exchange takes place are known both qualitatively (type of fluids, their approximate temperature levels, their corrosion characteristics, etc.) and quantitatively (mass flow rates, specific heats, etc.);
i) the Design Goal is a certain temperature gain in one of the fluids ("product") at the expense of a temperature drop of the other ("fuel"). Both of these temperature differences have been previously computed via proper mass- and energy balances;
j) the explicit Design Variable is the overall coefficient of heat transfer between the two streams, U_{th}: this coefficient depends on the relevant design variables which are geometric (tube diameter, spacing, thickness, surface finishing, configuration), physical (materials and their compatibility, etc.) and fluid dynamic (boundary layer thickness, turbulence characteristics, etc.). So, an accurate simulation procedure must be available, to compute U_{th} for each different configuration choice.

6.5.2 - SOME THEORETICAL CONSIDERATIONS

The problem can be solved via a search of a presumably low-dimensional solution space: the number of design options is limited both by theoretical considerations (high pressure stream ought to be on the tube side, the number of tube passes is limited by approach temperature difference limitations, etc.) and by technological restrictions (tubes ought to be circular in shape, tube arrangement is either in-line or staggered, fins can be pre-optimised with good approximation, use of turbulators on the inside is restricted by pressure drop considerations, etc.). In conclusion, the possible choices are:

7. stream allocation (hot stream on the tube side or on the shell side) → 2 possible alternatives
8. type of front head (single chamber mixed, split chamber mixed, single chamber unmixed, split chamber unmixed) → 4 possible alternatives
9. type of shell (vertical, horizontal, single pass, double pass) → 4 possible alternatives

10. type of rear head (U-tube, single chamber mixed, split chamber mixed, single chamber unmixed, split chamber unmixed) → 5 possible alternatives
11. number of tube passes (one, two, three... N) → N possible alternatives
12. tube arrangement (in-line or staggered) → 2 possible alternatives
13. tube diameter → M possible alternatives, depending on the commercially available tubes
14. tube type (unfinned, finned, turbulators inside, no turbulators inside) → 4 possible alternatives

There would be other design variables, but they are in effect secondary, and can be neglected here. Thus, the raw solution space contains (2*4*4*5*N*2*M*4) = 1280*N*M configurations, and an extensive search would appear very cumbersome (= resource consuming). But, there are as well some design constraints we can make use of:

4. high pressure stream should be on the tube side
5. fouling-prone stream should be on the shell side
6. hazardous stream should be on the tube side
7. if front head is mixed, so is usually the rear head; if front is unmixed, rear is also unmixed
8. the number of passes in practical cases never exceeds three
9. the tube diameter is dictated by the maximum velocity inside of the tubes, which is a discriminating parameter and depends on the volumetric flow rate. The choice is limited, i.e., M ranges between 2 and 6 (one, two or three commercial diameters above or below the recommended one)

These constraints in effect limit the number of possible configurations quite drastically: constraints I, II and III in practice reduce the 2 possibilities given by a) to 1; constraint IV reduces the 5 alternatives of d) to 2 (U-tube or "like the front end"); constraint V implies that N = 3 at most, and constraint VI limits M to 6. So, the total number of the possible choices ranges from (1*4*4*2*1*2*1*4) = 256 (if both N=1 and M=1) to (1*4*4*2*3*2*6*4) = 4608. So, an exhaustive search is feasible. Its implementation is straightforward, and follows the usual algorithmic procedures outlined in Heat Exchanger Design Handbooks (see for example [61]).

But we are interested in a Knowledge Based approach, and not in brute-force extensive searches! So, we are after a procedure that can strongly prune the decision tree (the number of configurations) in a non-algorithmic fashion. This can be done as follows:

4. rank the relevant design variables in descending order of importance
5. select all the configurations which the first design variable can generate;
6. attach to each configuration a numerical figure of merit which expresses our degree of technical appreciation or distaste for that configuration;

7. proceed to the next design variable (in logical sense, from the root to the trunks of the decision tree), and repeat the procedure. Discard the paths which are less appealing;
8. proceed similarly to the twigs and twiglets of the tree. At the end, only few leaves on some twiglets will remain: these are the "preferred" configurations, on which a sensitivity analysis or a refined extensive search can be performed if necessary.

6.5.3 - SOLUTION

To be able to develop the complete decision tree for the choice of the heat exchanger, we have reduced the number of possible alternatives given by the list presented in Section 6.5.2. This by no means detracts to the generality of the proposed procedure: in practical cases, there could be a "preliminary pruning" action executed by -or with the help of- a Domain Expert, with the goal of strongly reducing the viable alternatives, so that the decision tree results more manageable. The "reduced list" is thus the following:

8. stream allocation (hot stream on the tube side or on the shell side) → 2 possible alternatives
9. type of front → 2 possible alternatives head (either single chamber mixed or single chamber unmixed)
10. type of shell → 2 possible alternatives (single pass, double pass)
11. number of tube passes → 3 possible alternatives (one- , two- or N- passes)
12. tube arrangement → 2 possible alternatives (in-line or staggered)
13. tube diameter → 3 commercially available tubes
14. tube side aspect ratio → 3 possible alternatives ($ar =$ *tubespacing* $= 0.5$ or 1 or 2)

We shall rank the variables in the following descending order: stream allocation, shell type, tube passes, tube arrangement, tube diameter, tube aspect ratio, front head. In addition, the figure of merit will be the total (installation and operation) cost of a "unit heat exchanger" of identical type as that which represents the configuration emerging from our choices. Such a Unit Heat Exchanger is defined as the heat exchanger that performs the heat transfer under the same inlet- and outlet temperatures, but for a unit mass flow rate of the stream with the higher heat capacity. This is not an absolute criterion, and it has its shortcomings, but we will neglect these considerations by now, and assume that the convenience of this criterion has been proven by past experience. Then, after the first choice, the "tree" will look like Fig. 6.11a: notice that the figure of merit is expressed in arbitrary dimensionless units and is intended exclusively as an example of the applicability of the method. After the second variable is considered, the tree has 4 branches (Fig. 6.11b), two of which (2 and 4) are plagued by such high values of the penalty function that they can be discarded.

The procedure is repeated at a third, fourth...seventh level, and in the end, only the 3 configurations displayed in Fig 6.11c are left: on these, one can perform a detailed study.

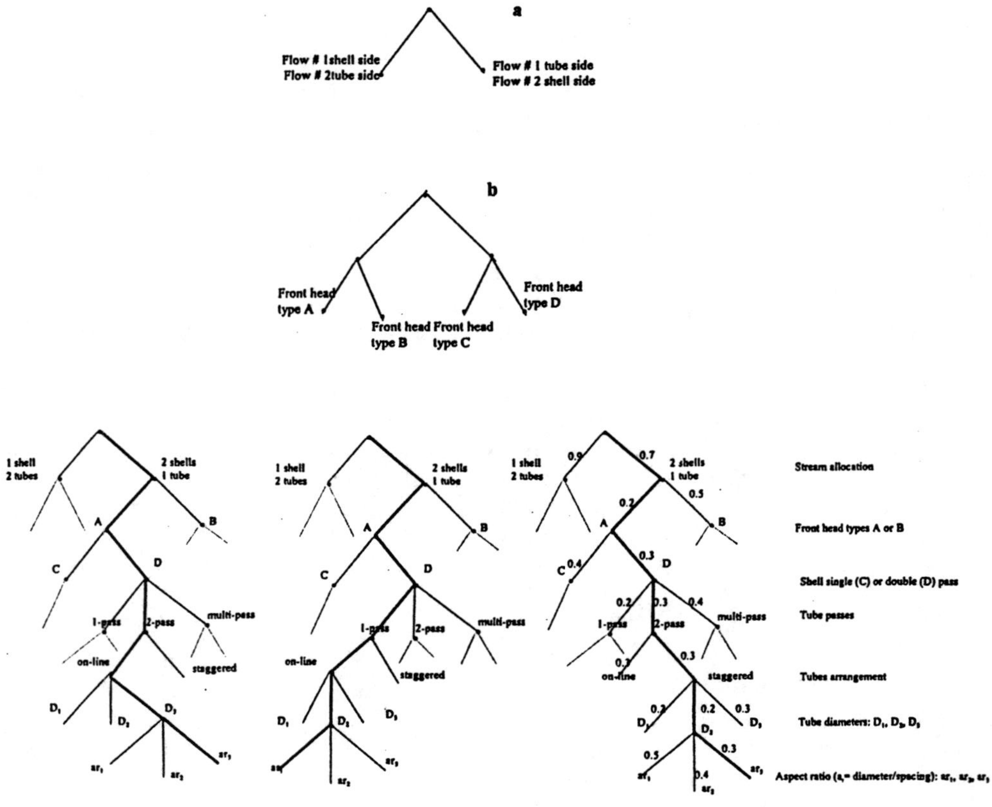

Figure 6-11. Decision tree for a heat exchanger selection

It is clear that the essential point here is to select the proper penalty function and to make sure that we do not leave out branches that would start with high penalty values and then in the end possess a low global penalty function: the selection of the penalty function is a highly specialised action and requires Domain Expert knowledge. In case of doubt, the procedure ought to be repeated with a slightly different penalty function formulation, to get an idea of the sensitivity of the solution to an error in the assessment of the operating costs.

6.6 - Choice and Technical Specification of a Gas Turbine Plant

6.6.1 - The physical problem

It is required that an automatic procedure be devised for the choice and specification of an open-cycle gas turbine plant. The procedure shall read the technical requirements (design data), choose the proper type of plant from a library of available plants, suggest possible (limited) cycle modifications, and produce a technical specification for the design engineer.

The following is a "task list" which can be constructed by consulting an energy system design textbook (or by asking an expert!):

- read design data: P_{el}, P_{therm}
- read constraints (if applicable): speed, service, fuel type(s), availability of process water for injection into the combustor, etc.
- check whether P_{therm}/P_{el} is within a feasible range
- scan library seeking for components (i.e., gas-turbine plants) with matching or *almost* matching P_{el}
- try to match type of service, fuel, rpm range, etc. with the problem data
- match P_{therm}/P_{el} of the proposed plant with design requirements
- produce a technical specification form

6.6.2 - Some theoretical considerations

This is still a selection problem, and it has been included here to show how to handle the very intensive calculations required by an even preliminary process simulation, by keeping them separate from the qualitative rules that guide the selection. The premises here are similar as those for the pump selection task: gas turbine technology is mature, and there are on the market only few commercially available designs, so the selection process reduces in effect to the scanning of a (small!) decision tree. It has to be noticed that it is the two remarks made above that make this reduction possible (i.e., “gas turbine technology is mature” and “there are only few commercially available configurations on the market”), and both are domain-specific knowledge that can be gathered only by consulting a Domain Expert.

6.6.3 - Solution

The flow chart of the procedure is depicted in Fig. 6.12. We will try to reduce the tree to a sequel of rules of the type:

IF p THEN q

where **p** is the *premise* (which can consist of more than one rule), and **q** is a *consequence*. Notice that "premise" and "consequence" are functionally equivalent to "cause" and "effect" here.

In this example, like in 6.4 above, the decision tree shows that, in addition to conditional rules, the procedure includes other steps which imply some *quantitative* decision, like "*simulate process, check P_j, etc.*". These steps will be treated as functional calls to some external numerical procedure, which will return the values of the properly addressed variables. Keeping this in mind, a set of rules which reproduces the decision tree of Fig.6.12 is the following:

rule #1): **IF** $P_{therm} / P_{el} \leq 1.1$
THEN abort procedure **AND** display warning
ELSE check library by P_{el}

rule #2): **IF** $0.9\, P_{el} \leq P_j \leq P_{el}$
THEN make suggestions for matching P_j and P_{el}
AND check library by $P_{el} = P_j$ and type of service

rule #3): **IF** available fuels match required fuel
THEN check whether $(P_{therm} / P_{el})_{library} \cong (P_{therm} / P_{el})_{required}$

rule #4): **IF** check is negative (i.e., no match)
THEN call GTSIM, change T_{max}, m_w to obtain match (iterate)
ELSE write technical specification

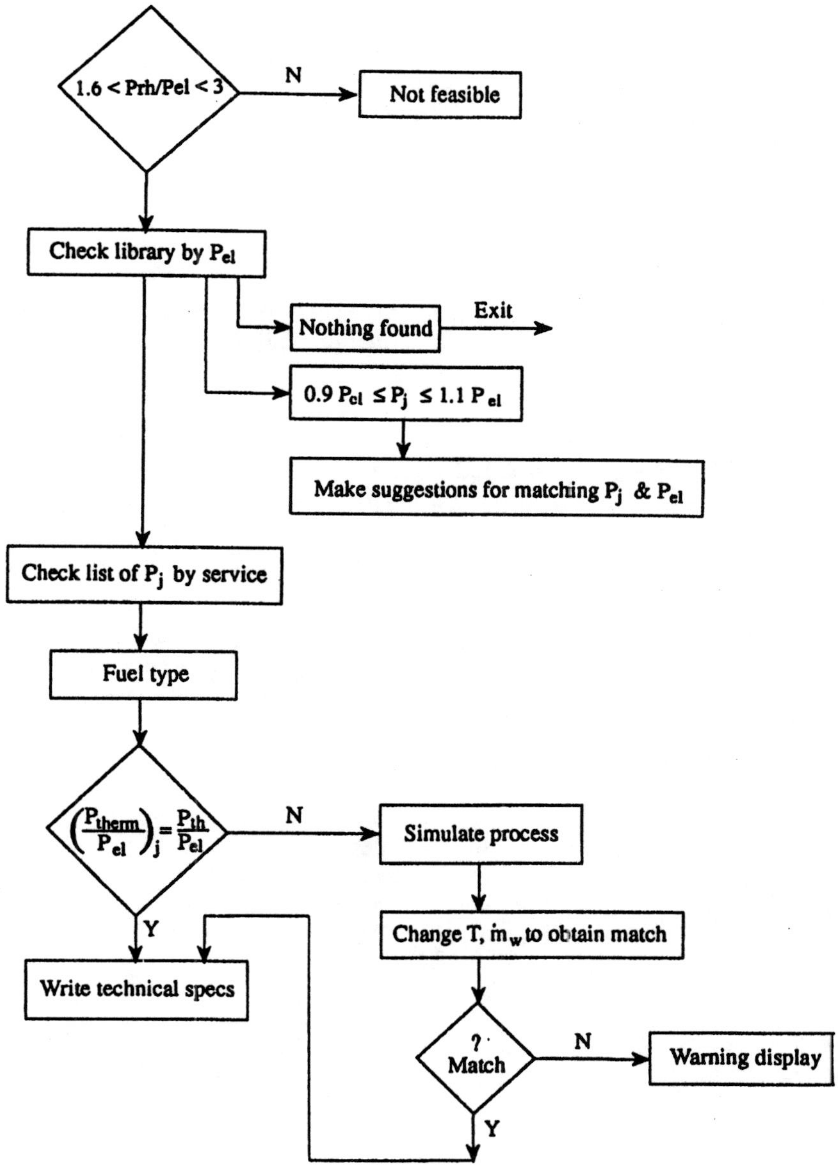

Figure 6-12. Decision tree for the choice of a gas-turbine plant

The remaining steps can be executed conditionally at the user's prompt.

As an example, the "suggestions" invoked by rule #2 to force a match between P_j and P_{el} could be derived as follows.

1) in a small neighbourhood of the design point, a variation in the output power is linearly proportional to a variation of T_{max} and of m_a:

$$\partial T \; {}^{-}\; \nu$$

2) the matching requires then that, (ΔP being the mismatch):

$$\Delta P = \alpha \, \Delta m_a + \beta \, \Delta T$$

which is a linear optimisation problem in two independent variables and requires an additional condition to be solvable. This additional condition can be a cost criterion, or an efficiency optimisation condition, etc.

6.7 - PROCESS DESIGN: CHOICE AND TECHNICAL SPECIFICATION OF A FOSSIL FUELLED POWERPLANT

6.7.1 - THE PHYSICAL PROBLEM

Consider the most general problem which a designer can be faced with in the field of "energy systems": that of *choosing the most appropriate configuration to extract an assigned amount of energy from a given resource base, under some "environmental" as well as "configurational" constraints*. The problem is formally posed in the following manner:

A) Design data:

1. the required power, which must be specified both in its type (electrical, mechanical, thermal) and in its *amount* (MW installed);
    1- the type of *environment*. This comprises the following information:
    - type and amount of available resources (fuel, air, other materials which can take part to the process);
    - amount of water available (m^3/sec);
    - any environmental characteristic which can be foreseen to influence the choice of the process (e.g.: "station is mounted on a barge", or similar)
    - all legal and/or technical constraints known to apply at the site under consideration.

B) A set of negative instances, i.e., of processes which cannot be -for whatever reasons- accepted as a solution (e.g.: "nuclear plant not acceptable").

6.7.2 - SOME THEORETICAL CONSIDERATIONS

This problem, apparently very complex, turns out to admit a relatively simple *qualitative* solution, but its complexity makes a *quantitative* solution difficult. We will present in this example a procedure (called "COLOMBO" from "*CO*nstrained, *LO*gically *M*odified *B*oundaries *O*ptimisation") which has been developed by the Authors [5], has been tested in several different realistic cases, and always found to be capable of completely solving the problem, robust and effective. A working copy of the procedure (executable on a UNIX system) is included -together with a brief users' manual- in the diskette attached to this book. The reader is encouraged to carefully read through this example and then try to test the procedure under different sets of "environmental conditions" (in the sense defined here below): the list of *possible plant configurations* produced by the code sheds light on the *way an expert system mimics the "thinking patterns" of a process engineer.*

First, consider that the Knowledge Base must include the technical, logical and physical description of the possible building blocks of such a plant, its components. At present, the procedure we are going to describe possesses its own default library of 23 components):

back-pressure steam turbine
combustion chamber
compressor
concentrated pressure loss
condenser
condensing steam turbine
contact (mixing) heat exchanger
deaerator
distributed pressure loss
electric generator
feedwater heater
flow merger
flow splitter
gas turbine
hydrostatic head
mechanical generator
power merger
power splitter
pump
steam generator
surface heat exchanger
throttling valve
waste-heat boiler

New components can be added. A component is identified by a set of input and output streams (specifying what kind of fluid it works on, what type of power it adsorbs or delivers, and identifying which output streams are the "main" streams and which one are the "by-products"), by a set of "design parameters" which uniquely identify it (inlet and outlet pressure, temperature, mass flow rate, power, etc.), and by a set of mathematical rules which completely specify both the inlet and outlet streams given certain proper sub-sets of the design parameters. A specific component is, in AI terms, an instance of a *class*; this class can therefore be divided into *sub-classes* according to some peculiar characteristic of each sub-class: here, the *power range* is chosen: so, the class "gas turbines" can be considered as comprising, say, four subclasses: "less than 5

MW", "between 5 and 30 MW", "between 30 and 100 MW", and "above 100 MW". These subclasses inherit of course all properties of the class they belong to, but can have different values (or ranges) of the design parameters. Each sub-class is further divided (where feasible) into three sub-sub-classes: "high tech", "state-of-the-art" ("standard"), and "low tech". Again, this allows for different values of specific design parameters (in the case of a gas turbine, for instance, inlet gas temperature ranges will differ).

6.7.3 - SOLUTION

The solution paradigm chosen here is an instance of the general procedure outlined in Section 5.6. It basically consists in a *strongly pruned, recursive scanning of a decision tree which is not known a priori*. The recursivity is required for two reasons: first, since the paths are not known in advance, there is the logical necessity of a quite extensive backtracking; second, a real process invariably possesses *physical loops*, which correspond to iterations in the logical build-up of the system. The procedure works *backwards*, considering the design goal as an effect, and trying to find its cause(s). Once the program is launched, since the design goal is given, the Inferential Engine scans the Component Library to seek the components which would deliver the goal (which is a specified amount of thermal and/or mechanical and/or electrical power). The first component which satisfies the goal is placed in the first slot of the working memory as the *last* component of a possible process P_1 which is still to be identified. If n components are found, there will be n possible (albeit yet unknown) "process trees" which the procedure has to work on. Then, for each P_i $(i=1,\dots n)$, the inputs required by the component just found are taken to be the new design goal, and the library is scanned again seeking for components whose outputs match the goal. The procedure is repeated until:

1. all the required inputs are available from the class "environment" and the configuration developed so far meets all specified constraints: in this case, a process has been found, and is displayed to the user in a proper form; (if more processes have been found, *all of them* are displayed;
a) at a certain level in the process tree, no match can be found (under the specified constraints) for the required inputs: the procedure is aborted.

The above is in practice an application of the so-called Heuristic-Increment-Generate-Test ("HGT", see for instance [74]), which amounts to an early pruning of inconsistent or highly unlikely solutions, to help reduce the resources allocated to the search algorithm. To "match" a design goal means that the output of the component under consideration must be of the same type of the design goal (mechanical power, etc.) and that the range attached to that sub-class shares at least a sub-range with the one required by the design goal. There are cases in which only the first of these two conditions is met, i.e., in which the output range of the component under consideration does not match the required input: in these cases, the component is added to the process, but the mismatch is recorded, for future action. In fact, the procedure at each step checks whether there is the possibility that one or more "by-products", or "secondary streams" can be used to force the match (for instance, increasing the temperature of the water flowing out of the condenser to match the boiler inlet temperature).

A "relative cost" is attached to each of the constructed processes: this cost is a function of the number of components and of the number and amount of external resources used. So, for instance, a steam powerplant with an externally fuelled feedwater heater train will have a higher "cost" than the same process with regenerative feedwater heating. COLOMBO displays the solution in the form of an interconnection matrix, M, in which the rows and the columns represent respectively components and streams: the elements of M can be:

$M_{i,j} = 0 \Rightarrow$ stream j does not interact with component i
$M_{i,j} = 1 \Rightarrow$ stream j is an input of component i
$M_{i,j} = -1 \Rightarrow$ stream j is an output of component i

The procedure can be directly "tested" by running it: here, we give a detailed description of its structure by discussing a specific example of its application. Consider the following problem:

1) find the most appropriate plant of nameplate power P = 100 MW, if P has to be delivered under the form of electrical energy;
2) the environment is thus composed:
   - fuel: natural gas, of given chemical composition, with LHV = 41000 kJ/kg, max. available mass flow rate m = 50 kg/s;
   - air, of standard chemical composition, with T_o = 32 C, relative humidity H_r = 5 %;
   - no water available
3) the following constraints apply:
   - plant factor P_f= 0.75 (6600 hours/year equivalent at nameplate load)
   - minimum conversion efficiency $\eta_{I,process} = 0.3$

These data have to be imparted to the code in a proper form: COLOMBO needs a mix of numerical values (m_w = 0, T_o = 32 C, etc.) and propositional strings ("plant_output_is_electrical"). The specific format of the database can vary from case to case, and is more a matter of convenience than an issue of substance. The "logical" steps of the inference are then the following:

1) *action*: scan components library ("CL") to find suitable components
   result: electrical generator; input: shaft work, approx. 102 MW
2) *action*: new goal 1 = shaft work, P_s = 102 MW
3) *action*: scan components library ("CL") to find suitable components
   result: gas-turbine high-tech; input: combustion-gas at $T_{g,max}$ = 1500 C

gas-turbine standard; input: combustion-gas at $T_{g,max} = 1300$ C
gas-turbine low-tech; input: combustion-gas at $T_{g,max} = 1100$ C
steam-turbine high-tech; input: steam at $T_{s,max} = 650$ C
steam-turbine standard; input: steam at $T_{s,max} = 550$ C
steam-turbine low-tech; input: steam at $T_{s,max} = 450$ C
diesel-engine; input: fuel-oil, m_{oil} 6 kg/s
power-splitter; input: shaft work

[We will not follow the "power-splitter" branch by now but, as we will see, this branch is also capable of generating valid configurations]

4) *action*: check constraints and environment
result: discard diesel-engine (no fuel oil in environment)

5) *action*: new goal 2 = hot pressurised gas at $T_{g,max} = 1500$ C
new goal 3 = hot pressurised gas at $T_{g,max} = 1300$ C
new goal 4 = hot pressurised gas at $T_{g,max} = 1100$ C
new goal 5 = hot pressurised steam at $T_{s,max} = 650$ C
new goal 6 = hot pressurised steam at $T_{s,max} = 550$ C
new goal 7 = hot pressurised steam at $T_{s,max} = 450$ C

6) *action*: scan components library ("CL") to find suitable components
result: for new goal 2: combustion-chamber high-tech; input: fuel and pressurised air, $p_{max} = 3$ MPa
result: for new goal 3: combustion-chamber high-tech; input: fuel and pressurised air, $p_{max} = 3$ MPa
result: for new goal 3: combustion-chamber standard; input: fuel and pressurised air, $p_{max} = 2$ MPa
result: for new goal 4: combustion-chamber high-tech; input: fuel and pressurised air, $p_{max} = 3$ MPa
result: for new goal 4: combustion-chamber standard; input: fuel and pressurised air, $p_{max} = 2$ MPa
result: for new goal 4: combustion-chamber low-tech; input: fuel and pressurised air, $p_{max} = 1.5$ MPa
result: for new goal 5: boiler high-tech; input: fuel, air and preheated water, $T_{fw,max} = 300$ C
result: for new goal 6: boiler standard; input: fuel, air and preheated water, $T_{fw,max} = 275$ C
result: for new goal 7: boiler low-tech; input: fuel, air and preheated water, $T_{fw,max} = 200$ C

[At this point, there are 6 "possible partial processes", and no constraint has been violated: for the sake of simplicity, we will follow goals 2 and 5, and neglect the standard and low-tech possibilities]

[Following the branch "new goal 2"]

7) *action*: new goal 8 = fuel
new goal 9 = pressurised air at $p_{max} = 3$ MPa

8) *action*: scan components library ("CL") to find suitable components

result: fuel available in "environment": branch search terminated
result: air compressor: input: air, shaft work

9) *action*: new goal 10 = air
new goal 11 = shaft-work

10) *action*: scan components library ("CL") to find suitable components
result: air available in "environment": branch search terminated
result: shaft-work is a by-product of gas-turbine
result: shaft-work is a main product of gas-turbine
result: shaft-work is a main product of power-splitter
result: shaft-work is a by-product of power-splitter

[Considering the shaft-work as a gas-turbine by-product]

11) *action*: try matching compressor required input with gas-turbine output
action: check constraints and environment: if no constraint is violated and all inputs come from the environment, the produced configuration is a legal one, and is displayed. A new branch is explored.

[Considering the shaft-work as a gas-turbine main product]

12) *action*: go to 5)

[Considering the shaft-work as a power-splitter main product]

13) *action*: go to 2)

[There is a built-in rule which does not allow a power splitter to be fed by another power splitter: this avoids looping]

[Considering the shaft-work as a power-splitter by-product]

14) *action*: go to 2)
[... and so on: there can be an infinite series of ever more complicated plants in which the compressor is powered by a gas-turbine plant whose compressor is in turn powered by yet another gas turbine plant: there are some specific rules in the IE to avoid excessive recursivity]

[Following the branch "new goal 5"]

15) *action*: new goal 12 = air
new goal 13 = fuel
new goal 14 = preheated water

action: scan components library ("CL") to find suitable components
result: air available in "environment": branch search terminated

result: fuel available in "environment": branch search terminated
result: deaerator; input: water, medium-pressure steam
16) *action*: new goal 15 = water
new goal 16 = medium-pressure steam
17) *action*: scan components library ("CL") to find suitable components
result: water not available in "environment": branch search failure
[medium-pressure steam is available from back-pressure-steam-turbine, water is also available from condenser]
result: condenser; input: low-pressure steam, water result: back-pressure-steam-turbine; input: high-pressure steam
18) *action*: new goal 17 = water
new goal 18 = low-pressure-steam
19) *action*: scan components library ("CL") to find suitable components
result: water not available in "environment": branch search failure
[low-pressure steam is available from condensing-steam-turbine; "water" for the condenser cannot be extracted from another condenser (to avoid looping)]
result: condensing-steam-turbine: input: medium-pressure-steam
20) *action*: go to 16), goal 16

["medium-pressure-steam" can also be obtained as a by-product of "steam-turbine", but this case (which would not modify the final outcome) is not considered here]

[This branch fails, because there is a recursive call to "medium-pressure-steam" which is terminated by an anti-looping rule built into the inference engine. It is important to notice that the physical reason of the failure is the absence of water in the environment, which negates the insertion of a "condenser" into the process and the construction of a "steam plant" configuration]
21) *action*: check if there are any unexplored branches. If there is none, terminate the run.

A list of the first four configurations produced by the above run is shown in fig. 6.13 through 6.17. Notice that configuration 1 has been produced because, since electricity is available from the net, the code has assumed -wrongly- that it can drive a 100 MW-electrical motor directly from the net (see example 2 in Section 3.3).

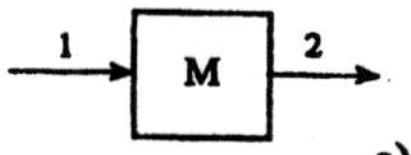

a)

item \ flow	1 work	2 work
ELECTRIC MOTOR	1	-1

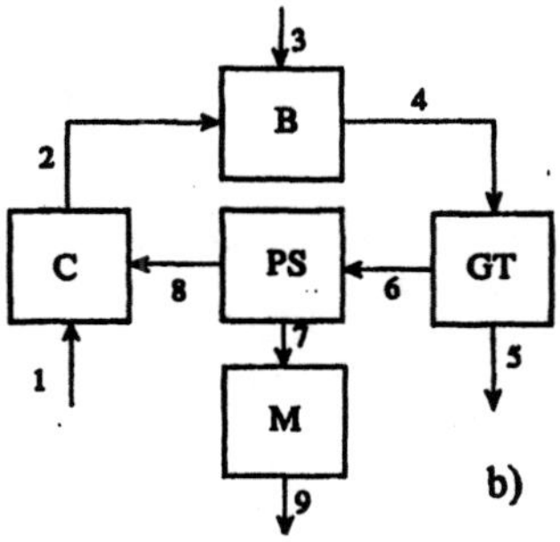

b)

item \ flow	1 air	2 air	3 fuel	4 hot gas	5 hot gas	6 work	7 work	8 work	9 work
COMPRESSOR	[illegible]	[illegible]	0	0	0	0	0	[illegible]	0
BOILER	0	[illegible]	[illegible]	[illegible]	0	0	0	0	0
GAS TURBINE	0	0	0	[illegible]	[illegible]	[illegible]	0	0	0
POWER SPLITTER	0	0	0	0	0	[illegible]	[illegible]	-1	0
ELECTRIC MOTOR	0	0	0	0	0	0	[illegible]	0	[illegible]

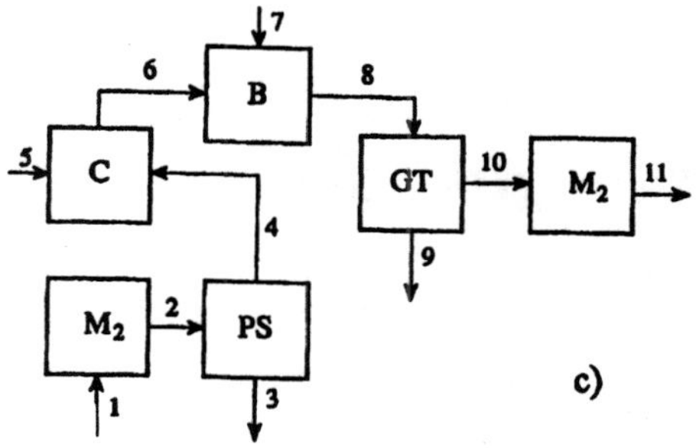

c)

item \ flow	1 elect.	2 work	3 work	4 work	5 air	6 air	7 fuel	8 hot gas	9 hot gas	10 work	11 work
ELECTRIC MOTOR 1	[illegible]	-1	0	0	0	0	0	0	0	0	0
POWER SPLITTER	0	[illegible]	-1	-1	0	0	0	0	0	0	0
COMPRESSOR	0	0	0	1	1	-1	0	0	0	0	0
BURNER	0	0	0	0	0	[illegible]	1	[illegible]	0	0	0
GAS TURBINE	0	0	0	0	0	0	0	1	-1	-1	0
ELECTRIC MOTOR 2	0	0	0	0	0	0	0	0	0	1	-1

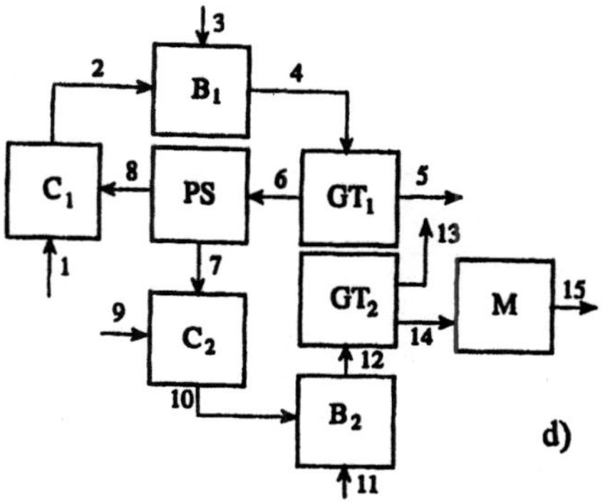

d)

item \ flow	1 air	2 air	3 fuel	4 hot gas	5 hot gas	6 work	7 work	8 work	9 air	10 air	11 fuel	12 hot gas	13 hot gas	14 work	15 work
COMPRESSOR 1	[illegible]	[illegible]	0	0	0	0	0	[illegible]	0	0	0	0	0	0	0
BURNER 1	0	[illegible]	[illegible]	[illegible]	0	0	0	0	0	0	0	0	0	0	0
GAS TURBINE 1	0	0	0	[illegible]	[illegible]	[illegible]	0	0	0	0	0	0	0	0	0
POWER SPLITTER	0	0	0	0	0	[illegible]	[illegible]	[illegible]	0	0	0	0	0	0	0
COMPRESSOR 2	0	0	0	0	0	0	0	[illegible]	[illegible]	-1	0	0	0	0	0
BURNER 2	0	0	0	0	0	0	0	0	0	[illegible]	[illegible]	-1	0	0	0
GAS TURBINE 2	0	0	0	0	0	0	0	0	0	0	0	1	-1	-1	0
ELECTRIC MOTOR	0	0	0	0	0	0	0	0	0	0	0	0	0	1	-1

Figure 6-13. The first four process configuration suggested by COLOMBO for a power generation process

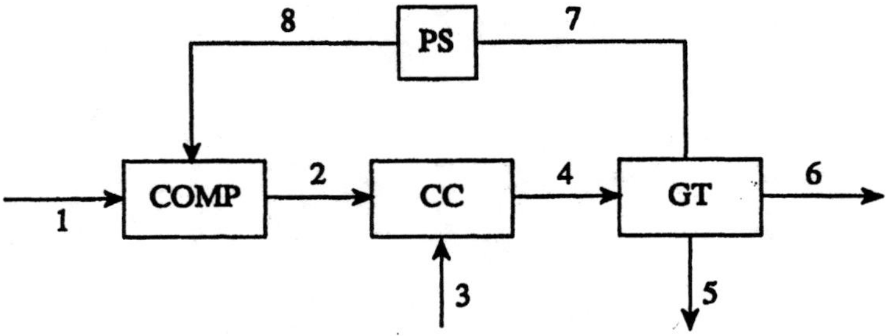

item \ flow	1 air	2 air	3 fuel	4 hot gas	5 hot gas	6 work	7 work	8 work
COMPRESSOR	1	-1	0	0	0	0	0	1
BURNER	0	1	1	-1	0	0	0	0
GAS TURBINE	0	0	0	1	-1	-1	-1	0
POWER SPLITTER	0	0	0	0	1	0	1	-1

Figure 6-14. The first process resulting from the "GAS-TURBINE" choice

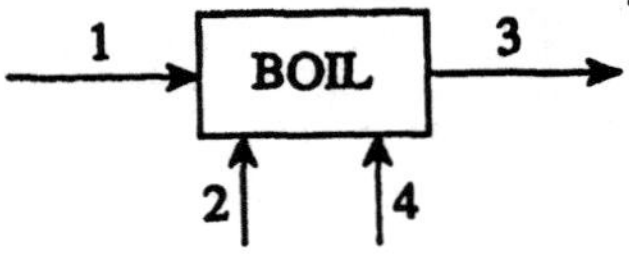

item \ flow	1 air	2 fuel	3 water	4 steam
BOILER	1	1	1	1

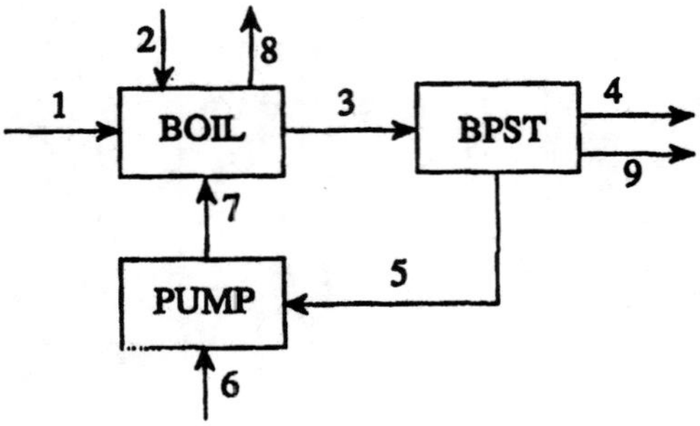

item \ flow	1 fuel	2 air	3 steam	4 work	5 work	6 water	7 water	8 exhaust	9 steam
BOILER	1	1	-1	0	0	0	1	-1	0
BP STEAM	0	0	1	-1	-1	0	0	0	-1
PUMP	0	0	0	0	1	1	-1	0	0

Figure 6-15. The first two processes resulting from the "STEAM" choice

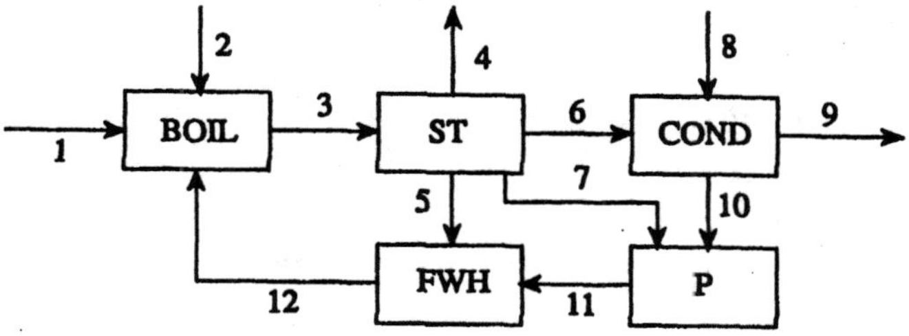

item \ flow	1 air	2 fuel	3 hot gas	4 work	5 steam	6 steam	7 water	8 cool. water	9 cool. water	10 water	11 water	12 water
BOILER	[illegible]	[illegible]	[illegible]	0	0	0	0	0	0	0	[illegible]	0
STEAM TURBINE	0	0	[illegible]	[illegible]	[illegible]	[illegible]	[illegible]	0	0	0	0	0
CONDENSER	0	0	0	0	0	[illegible]	0	[illegible]	[illegible]	[illegible]	0	0
PUMP	0	0	0	0	0	0	0	0	0	[illegible]	[illegible]	0
FEED WATER HTR	0	0	0	0	[illegible]	0	0	0	0	0	[illegible]	[illegible]

Figure 6-16. The first process resulting from the "STEAM-TURBINE" choice

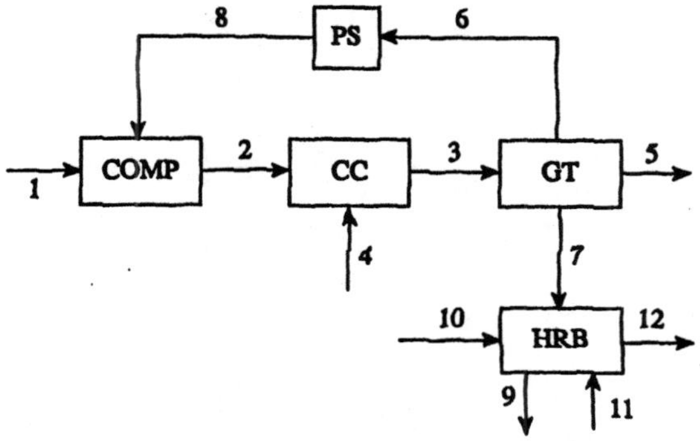

item \ flow	1 air	2 air	3 hot gas	4 fuel	5 work	6 work	7 hot gas	8 work	9 hot gas	10 water	11 fuel	12 steam
COMPRESSOR	[illegible]	[illegible]	0	0	0	0	0	[illegible]	0	0	0	0
BURNER	0	[illegible]	[illegible]	[illegible]	0	0	0	0	0	0	0	0
GAS TURBINE	0	0	[illegible]	0	[illegible]	[illegible]	[illegible]	0	0	0	0	0
POWER SPLITTER	0	0	0	0	0	[illegible]	0	[illegible]	0	0	0	0
HEAT REC. BOILER	0	0	0	0	0	[illegible]	0	0	[illegible]	[illegible]	[illegible]	[illegible]

Figure 6-17(a). The first process resulting from the "POWER" and "STEAM" choice

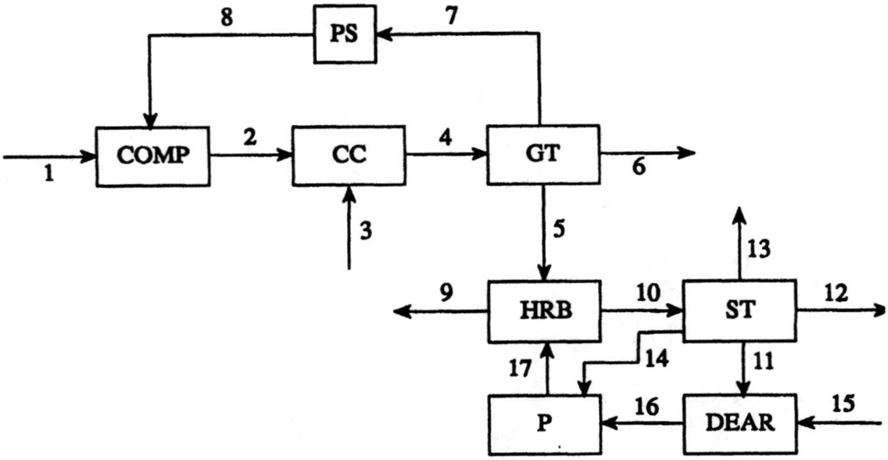

item \ flow	1 air	2 air	3 fuel	4 hot gas	5 hot gas	6 work	7 work	8 work	9 hot gas	10 steam	11 water	12 work	13 steam	14 water	15 steam	16 wwater	17 water
COMPPRESSOR	[illegible]	[illegible]	0	0	0	0	0	0	0	0	0	0	0	0	0	0	0
BURNER	0	[illegible]	[illegible]	[illegible]	0	0	0	0	0	0	0	0	0	0	0	0	0
GAS TURBINE	0	0	0	[illegible]	[illegible]	[illegible]	[illegible]	0	0	0	0	0	0	0	0	0	0
POWER SPLITTER	0	0	0	0	0	0	[illegible]	[illegible]	0	0	0	0	0	0	0	0	0
HEAT REC. BOILER	0	0	0	0	1	0	0	0	[illegible]	[illegible]	0	0	0	0	0	0	[illegible]
STEAM TURBINE	0	0	0	0	0	0	0	0	0	[illegible]	[illegible]	[illegible]	[illegible]	[illegible]	0	0	0
DEAERATOR	0	0	0	0	0	0	0	0	0	0	[illegible]	0	0	0	[illegible]	[illegible]	0
PUMP	0	0	0	0	0	0	0	0	0	0	0	0	0	[illegible]	0	[illegible]	[illegible]

Figure 6-17b. The second process resulting from the "POWER" and "STEAM" choice

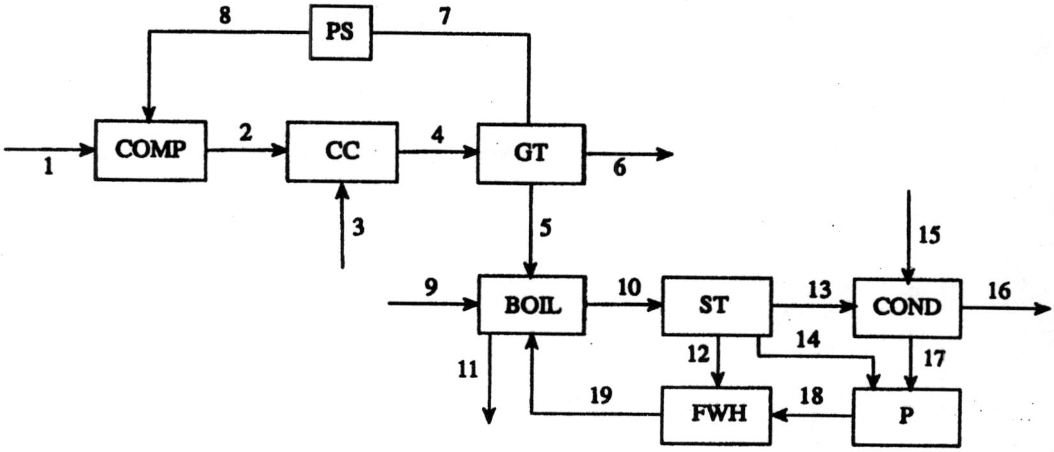

item \ flow	1 air	2 air	3 fuel	4 hot gas	5 hot gas	6 work	7 work	8 work	9 water	10 steam	11 hot gas	12 steam	13 steam	14 water	15 cool. water	16 cool. water	17 water	18 water	19 water
COMPRESSOR			0	0	0		0	0	0	0	0	0	0	0	0	0	0	0	0
BURNER	0				0	0	0	0	0	0	0	0	0	0	0	0	0	0	0
GAS TURBINE	0	0	0					0	0	0	0	0	0	0	0	0	0	0	0
POWER SPLITTER	0	0	0	0	0	0			0	0	0	0	0	0	0	0	0	0	0
BOILER	0	0	0	0		0	0	0				0	0	0	0	0	0	0	0
STEAM TURBINE	0	0	0	0	0	0	0	0	0		0				0	0	0	0	0
CONDENSER	0	0	0	0	0	0	0	0	0	0	0	0	0	0				0	0
PUMP	0	0	0	0	0	0	0	0	0	0	0	0	0		0	0			0
FEEDWATER HTR	0	0	0	0	0	0	0	0	0	0	0	0	0	0	0	0	0		

Figure 6-17c. The third process resulting from the "POWER" and "STEAM" choice

6.8 - Choice of "Optimal" Steam and Power Generation for Plant Expansion

6.8.1 - The Physical Problem

It is required that an automatic procedure be devised for the choice and specification of a process capable of producing additional power and process steam for a plant already in operation. The procedure shall read the technical requirements (design data), build-up the proper type of plant from a library of available components, and produce a technical specification for the design engineer.

Referring to Example 6.6, we will reformulate (and expand a little bit) the problem as follows:

- Design goal: devise a process capable of delivering power and process steam
- Design data: P 30 MW; $m_{steam} \geq 14$ kg/s
- Design constraints: $x_{steam} > 1$ (superheated steam)
  $P_{steam} \geq 2.$ MPa
  $NO_x \leq NO_{x,r}$
  $CO \geq CO_r$
  $SO_2 \geq SO_{2,r}$
  $O_2 \geq O_{2,r}$

(where the suffix "r" denotes the mandatory legal limit).

- Design environment: air, standard quality, unlimited quantity
  water, standard quality, unlimited quantity
  fuel, natural gas, $m_f = 10$ kg/s max

6.8.2 - Some Theoretical Considerations

This task is similar to the one examined in the previous example: the difference is the presence of some additional constraints, which allow for a "qualitative optimisation". The same code (COLOMBO) described above will be employed here.

6.8.3 - SOLUTION

The approach we will take in solving the problem is again backward-chaining induction which starts from the design data (symptoms) and will try to reach the design goal (diagnosis) while abiding to all constraints. The procedure needs some domain-specific engineering knowledge: for the present purposes, let us assume it possesses a sufficiently complete library L of engineering components, and one set of property tables for each working fluid. The inductive chain will work as follows (see also Example 6.7 above):

i) the Object POWER can be delivered by the following Objects contained in L:

1- GAS TURBINE
2- STEAM TURBINE
3-DIESEL ENGINE
4-ELECTRICAL MOTOR

ii) the Object STEAM can be delivered by the following Objects contained in L:

5- BOILER
6- BACK PRESSURE STEAM TURBINE

iii) the procedure now analyses each single possibility (thus originating a decision tree):

iv) the Object GAS TURBINE needs as input the following object:

7- COMPRESSED HOT GAS

which can be produced by:

8- PRESSURISED COMBUSTION CHAMBER
9- GASTURBINE

Taking the first path:

v) PRESSURISED COMBUSTION CHAMBER needs:

10- FUEL
11- COMPRESSED AIR

FUEL is available in the environment.

vi) COMPRESSED AIR needs:

12- COMPRESSOR

which in turn needs:

13- AIR
14- POWER

AIR is available in the environment.

vii) POWER needs:

15- GAS TURBINE
16- STEAM TURBINE
17-DIESEL ENGINE
18-ELECTRICAL MOTOR

Now, one of these Objects (the Gas Turbine) has been already selected, and the procedure makes use of a "priority rule":

rule #xx): **IF** last Object found belongs to a previously selected list
THEN terminate search **AND** place Object in CONFIGURATION LIST

This rule is applied every time an Object is found after a library scan. Another rule in the Inference Engine states that:

rule #yy): **IF** all inputs required by CONFIGURATION LIST are available in ENVIRONMENT
OR are outputs of Objects in CONFIGURATION LIST
THEN terminate branch **AND** store CONFIGURATION LIST in POSSIBLE PLANTS LIST
AND start searching next branch

We can see that the inductive chain has produced a path *from the design goal* to the environment and that this path represents the scheme of a process as shown in Fig. 6.18. Similarly, the search executed with the same methodology on the branch starting with STEAM TURBINE would produce the process schemes shown in Fig. 6.19. The other two branch searches would fail, because both DIESEL ENGINE and ELECTRICAL MOTOR need inputs that are not in the "environment" (Diesel oil and electricity respectively). The search which starts from the object STEAM results in the four processes shown in Fig. 6.20.

To solve the second part of the problem, i.e., to choose the optimal process, one must use *on each configuration proposed by the code* one of the optimisation techniques available in the literature [4,53,76].

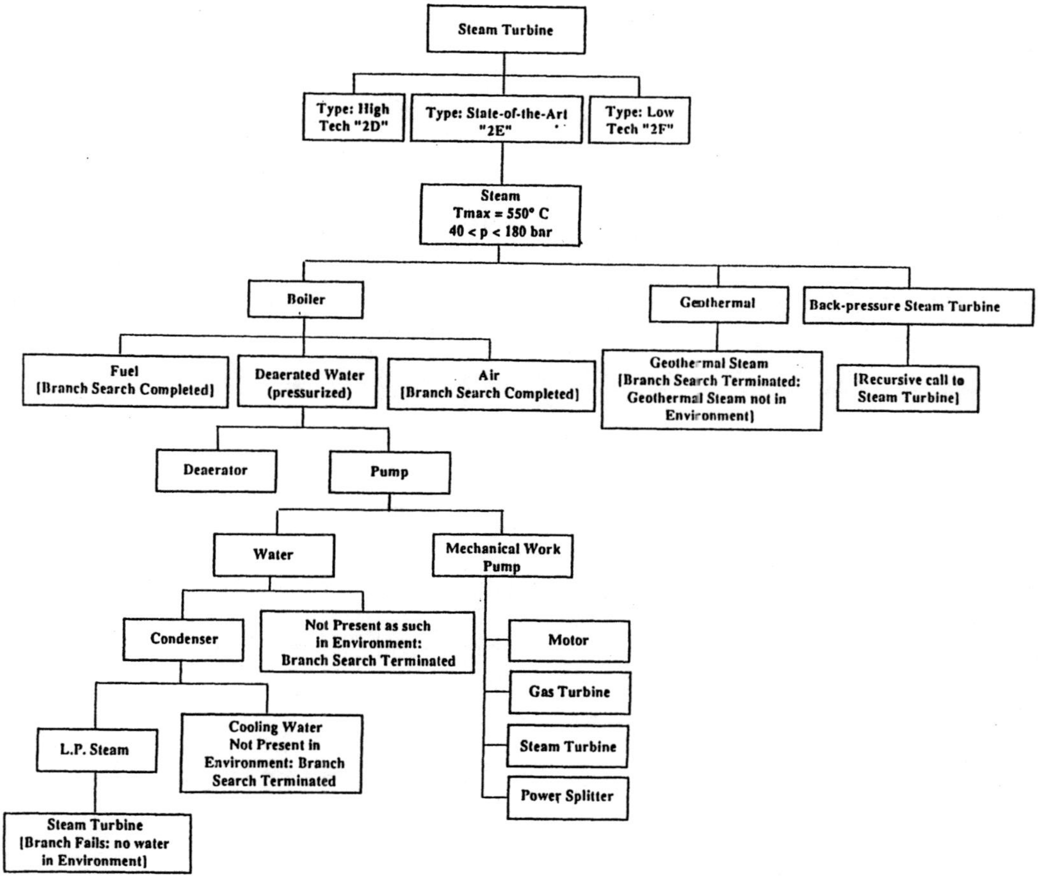

Figure 6-18. Decision tree for Steam Turbine Generation Process Selection and Specification

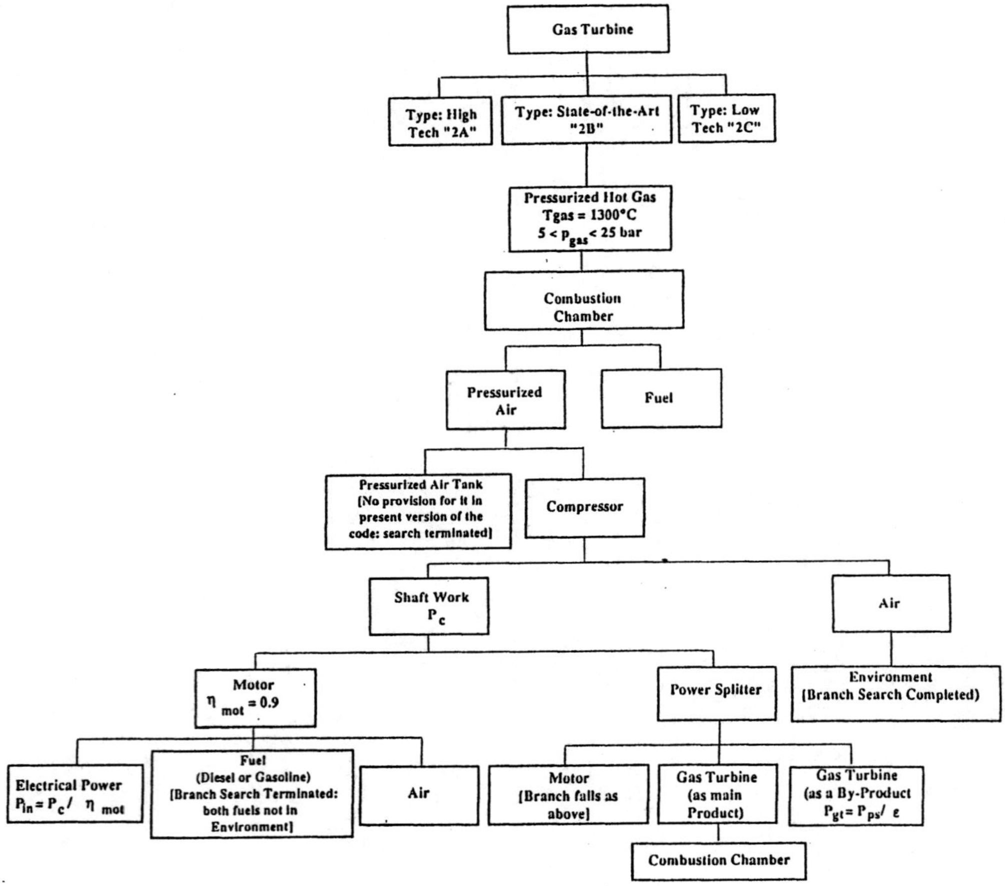

Figure 6-19. Decision Tree for Gas Turbine Generation Process Selection and Specification

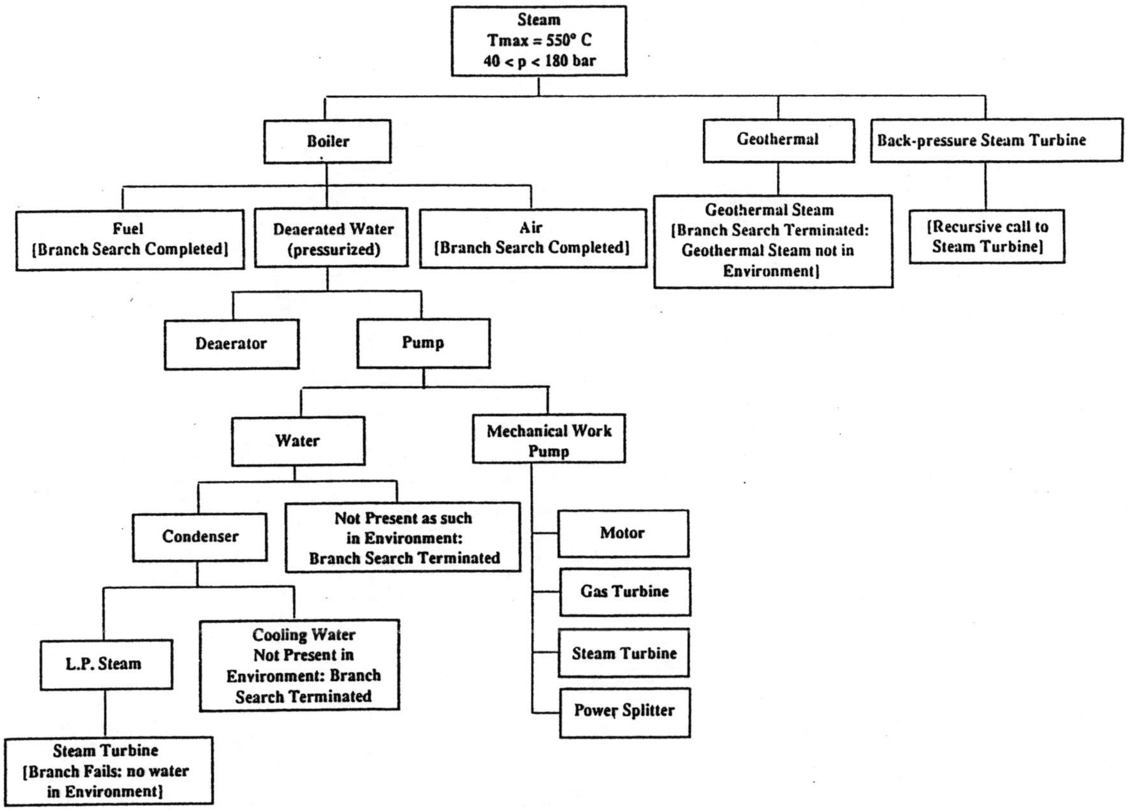

Fig. 6-20 Decision Tree for Steam Generation Process Selection and Specification

6.9 - OPTIMAL SYNTHESIS OF A HEAT EXCHANGER NETWORK

6.9.1 - THE GENERAL PROBLEM

"Heat Exchanger Network" ("HEN") is a common engineering denomination used to describe *an interconnected system of liquid or gaseous streams interacting in such a way that some of them gain a certain amount of energy in the form of sensible or latent heat at the expenses of some of the others.* Since the thermal exchange obviously takes place in heat exchangers, a HEN can also be seen as *a system of Heat Exchangers connected through their inputs and outputs.* Though the two descriptions are of course equivalent, the second one is more suitable for our purposes, and shall be adopted here. A typical HEN design problem is formulated as follows: given an industrial process that contains S streams that need heating or cooling, and a set of "targets", i.e. of specifications of the physical and thermodynamic conditions of each stream in each representative section of the process, devise a system of interconnected heat exchangers (HEN) such that:

3. the specifications are met
c) the HEN is "optimal" with respect to a well-defined objective function;
c) the direct and indirect constraints dictated by the main industrial process are abided by.

The heat donor streams are called "hot", the heat acceptors "cold". In addition, there may be "hot" and "cold" utilities, i.e., high temperature sources that may be used for heating up (some of) the cold streams, and heat sinks (usually, cooling water at ambient or near-ambient temperature) that may be used for cooling down (some of) the hot streams. One should consider that the problem invariably carries case-dependent complications that are not apparent in the elementary formulation presented above: some of the fluids may experience phase changes; most of the fluids have temperature-dependent physical properties; under certain circumstances mixing heat exchangers may be considered; the heat exchange coefficients vary with the type of flow; there are certain restrictions on the actual configuration of a heat exchanger, etc. The HEN design is therefore a very complex problem, and for more than 10 streams of each "type" it becomes intractable in practice. Since in today's industrial processes there may be as many as 100 streams interacting (and in most simple thermal processes, excluding refineries and chemical plants, it is common to find 30-40 streams-HENs), there is a high demand from Process Engineers for an "automatic" (in some sense) Design Procedure: though the problem is far from being solved, we shall try in this example to lay the guidelines for the implementation of an Expert Assistant for HEN Design.

6.9.2 - SOME THEORETICAL CONSIDERATIONS

The first approach which comes to mind is of course a "brute force" combinatorial method: the idea here is that of constructing and testing *all of the possible combinations* of heat exchangers that meet the mass- and energy specifications, and selecting the "optimal" one. This corresponds in practice to an exhaustive tree-scanning, and the practical implementation of such an algorithm would pose no problem. Unfortunately, the *size* of the problem makes the approach unfeasible: with N_h "hot" streams and N_c "cold" streams, the total number of possible and distinct HENs is $N_{HENs}= (N_h * N_c)!$: for 5 hot and 4 cold streams, 20!= 24329 10^{14} configurations ought to be tested![5] The *combinatorial* approach, even if adopted only for "small" problems, would thus lead to an extraordinary consumption of computer resources, and would therefore result very expensive: notice that for each configuration, a complete thermal energy balance must be carried out for each heat exchanger. Worst yet, it must be realised that any method based on a combinatorial technique is mathematically intractable, because the computational resources required by its solution grow more than exponentially. These two considerations seem to prohibit at all every combinatorial approach to the HEN Design problem: the solution tree becomes unwieldy. Fortunately, things change drastically if one introduces some heuristics in the search: this has been done by Hohmann [24] who used some energy-based arguments to derive a minimum number of Heat Exchangers in the HEN (see Section 6.9.3). This "minimum number rule" also limits indirectly the total number of systems to analyse in detail, because it is reasonable to explore only those configurations having a number of components in the vicinity of the minimum. Furthermore, one can make use of additional "Engineering Common Sense" considerations, that for instance prohibit the transfer of heat from a cold stream to a hot one, and limit the possible remaining combinations by declaring "feasible" only those that meet a specified restriction both on the *approach temperature difference*, "ATD", (difference between the incoming temperature of a cold stream and the outlet temperature of the corresponding hot stream, see figure 6.21) and on the *terminal temperature difference*, "TTD", (difference between the final temperature of a cold stream and the inlet temperature of the corresponding hot stream, see Fig. 6.21)[6]. This simple (qualitative!) criterion

[5] Notice that in reality the configurations to be tested in an exhaustive search are much more than N_{HENs}: if one considers incomplete pairings, i.e. configurations in which not all hot fluxes interact with all cold ones, the number of possible distinct configurations rises to $N_{tot}= n+n*(n-1)+n*(n-1)*(n-2)+...n!$, where $n = N_h * N_c$. With N_h=2 and N_c=2, N_{tot}= 64 (while N_{HENs}= 24); with N_h=5 and N_c=4, N_{tot}= 66133 10^{18};

[6] Both the approach ("ATD") and the terminal ("TTD") temperature differences are defined here for the "equivalent counterflow heat exchanger" (see [11]): if a parallel heat exchanger had been instead selected as reference, the ATD would have been the temperature difference between the incoming hot and cold streams, and the TTD that between the corresponding final conditions at exchanger outlet.

would produce a very strong pruning of the configuration tree, at the cost of some additional computational effort (both the ATD and the TTD must be calculated by preliminary energy balances). But, the number of alternative configurations to examine would still be high: for a (10+10) streams case, even if "stream splitting" is not enforced (see below, Section 6.9.3), about 100 different configurations must be designed and compared. Several methods have been proposed in the last three decades to solve the "Optimal HEN Problem" (for a comprehensive, though not very recent, review, see [23], and one of the most important outcomes of the rich and lively debate that is still continuing at present is that the very same definition of an "Optimal HEN" should be subject to scrutiny in the course of the problem formulation, because different choices (Minimal Area, Maximum Energy Recovery, Minimal Cost) can lead to the exploration of entirely different branches of the configuration tree. We will not try to generate an Expert HEN Designer (i.e., an ES that solves the synthesis-and-optimisation problem), and take a pragmatic approach instead: let us try to build an *Expert Assistant to the HEN designer* [75], that can guide the engineer in the search process through the configuration tree, and guarantee the attainment of at least a quasi-optimal configuration under the specified objective function. It is noteworthy that this Expert Assistant relies on a rather limited number of very powerful rules, instead of availing itself of a large number of specific rules of limited range: this derives from the fact that the particular set of rules adopted here has been "distilled" from the aggregate knowledge of generations of HEN Designers and Thermodynamicists, and therefore represents *deep* rather than *superficial* Procedural Knowledge.

6.9.3 - THE PHYSICAL PROBLEM FORMULATION

Consider S fluid streams, N_h of which are to be cooled and N_c heated to specified temperatures. We shall call here N_h the "hot" and N_c the "cold" streams. The following knowledge is also available:

1. all information concerning mass flow rates, initial temperatures, heat capacities and physical properties of each stream has been properly acquired;
2. the final temperatures of each stream ("target temperature") are specified by process requirements. If these targets are weak ("T_{final} of *K-th* stream must not be below 345°K"), or if they are process-dependent, a base configuration will be chosen for which the targets represent strong constraints;
3. both the availability and the cost of the "utilities", i.e., the process streams that may be used as heat sources and sinks for the HEN, are known. Furthermore, their cost structure is also known, so that variations in utility requirements may be correctly priced;
4. capital and operating costs of all equipment directly used in the HEN (pumps, valves, pipes, structures and heat exchangers), as well as marginal

capital- and operating costs of all equipment that can be indirectly affected by the HEN performance are known;

5. heat exchange coefficients are assigned. Since they depend on the individual Heat Exchanger configuration, only approximate averaged values are available at the time of problem formulation: in the course of the design, these initial values may be recalculated with closer approximation.

The solution consists in choosing the optimal HEN configuration, according to a specified objective function. Consider the process shown in Fig. 6.22 (this example has been adapted from [1]): there are 2 cold- and 2 hot streams, and therefore the extension of the purely combinatorial problem would be $N_{HENs}= 24$, or $N_{tot}= 64$. In reality, what is of interest here is that it is the *total number of heat exchangers* that is limited by a general rule derived by Hohmann [23]: assuming that it is always possible to obtain the desired heat exchange between two streams in only one piece of equipment[7], the *minimum number* of Heat Exchangers that will constitute the HEN is given by the so-called Hohmann formula:

$$N_{min} = N_h+N_c+ N_{hu}+N_{cu}-1$$

in which N_{hu} and N_{cu} represent the number of hot and cold utilities respectively, for each one of which a separate heat exchanger must be inserted[8]. A first rule in our KB is therefore:

rule 1): IF ($N_{he} > N_{min}$)
THEN (discard present configuration)

The reader will notice that there is a logical gap here: to apply rule#1, it is necessary that a "present configuration" already exists, i.e., that there is some assembly procedure that "produces" configurations. Without any loss of generality, we can stipulate here that any of the existing combinatorial methods may be used: we are guaranteed, by a correct application of the above rule, that only a small fraction of the "mathematically possible" configurations will be actually considered feasible by our Expert Assistant.

[7] Notice that this is a strong requirement, and it is not always possible to satisfy it in practice: for instance, several modular heat exchangers might be available, for a certain process, at substantial lower cost than a single, larger unit.

[8] Hohmann's rule admits exceptions, as it was pointed out by several researchers and subsequently experimentally demonstrated [64]. For simplicity, we shall instruct our Expert Assistant to ignore the exceptions.

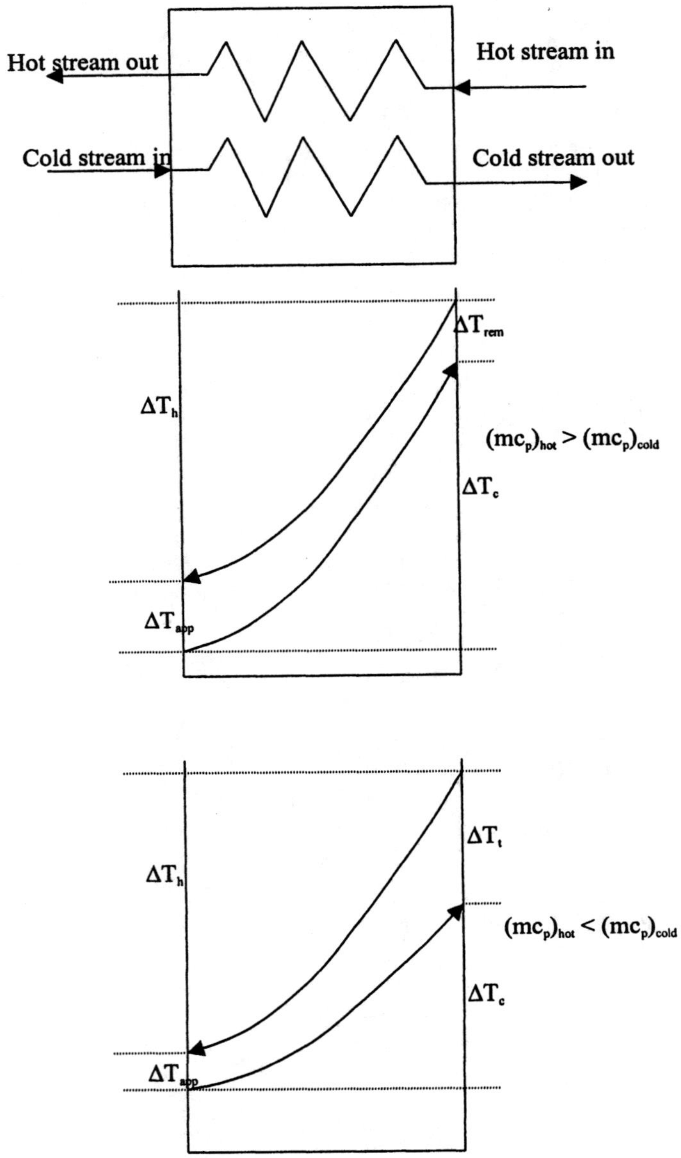

Figure 6-21. Temperature profiles in the equivalent counterflow heat exchanger

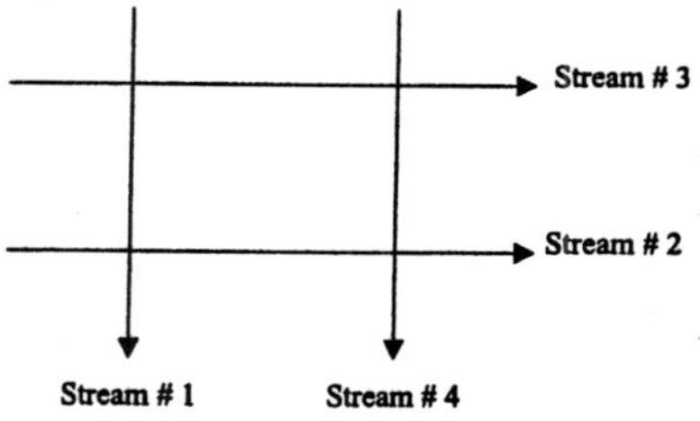

Stream Number	Inlet Temperature T_i (K)	Outlet Temperature T_o (K)	Heat Capacity Rate mc_p (kW/K)
1 (h)	400	310	2.0
2 (c)	300	390	1.8
3 (c)	330	370	4.0
4 (h)	450	350	1.0

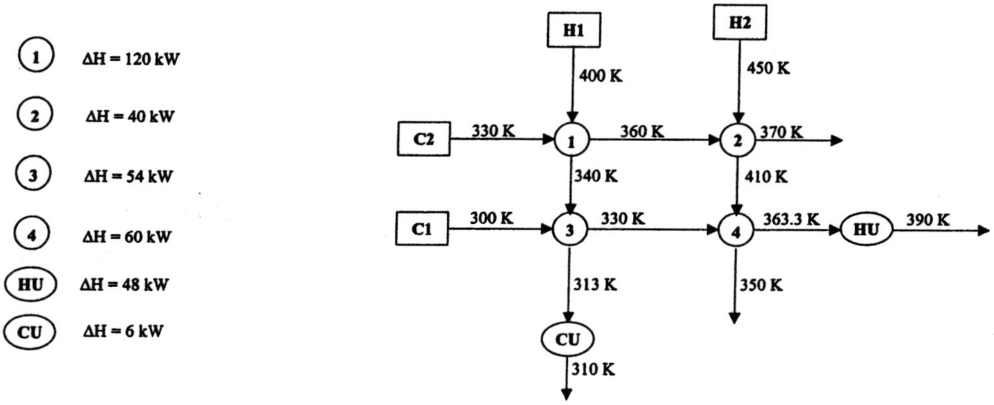

Figure 6-22. Benchmark 4SPI (Pho & Lapidus, as quoted in [1]): HEN configuration indicated as "optimal" by HENEA

Rule#1 can be considered as a heuristic rule: to proceed effectively in our path towards an optimal design, we can make use of other rules, equally powerful as Rule#1, that express a different kind of knowledge about the problem. These rules must be expressions of the "deep knowledge" that an expert human designer makes use of when attempting to solve the problem: several sets of guidelines have been compiled, and some have been actually implemented in (deterministic) numerical codes that apply them in a pre-defined order to attain a solution. These codes are Problem Solvers, in that their path towards a solution is not influenced at all, or only in minimal part, by a direct involvement on the part of their users[9]. What we are after here is though something different: we are trying to implement an Expert Assistant, whose task is not that of finding the solution in an autonomous way, but rather that of providing "suggestions" or "guidelines" to the user, who can accept the ES advice or make decisions on his own. The choice of the set of rules to implement is a matter of "taste": our experience in Thermal Systems Synthesis leads us to suggest using the so-called "common sense Second Law rules" suggested by [16], and later augmented by [1,77]. It is important to remark that pragmatism ought to guide the decision here, and that a different set of rules should be adopted if it can be demonstrated that it can lead to an equivalent solution with a smaller computational effort, or to a better result with the same computational effort. This said, the proposed set of rules is the following (list adapted from [16]:

1. do not use excessively large or excessively small thermodynamic driving forces in process operation
2. avoid whenever possible the mixing of streams; if mixing is unavoidable, avoid bringing in contact streams with substantial differences in temperature, pressure or chemical composition;
3. do not discard, directly or indirectly, heat at high temperature into the environment;
4. do not heat refrigerated streams with hot streams, or with hot- or cold utilities;
5. when choosing streams for heat exchange, try to match streams where the final temperature of one is close to the initial temperature of the other;
6. avoid whenever possible to exchange heat between streams with large difference in *flow* heat capacity (product of the mass flow rate with the heat capacity);
7. when exchanging heat between two streams, minimise the use of intermediate heat transfer fluids;
8. rank the heat content of the streams: heat (and refrigeration)
9. more valuable, the further their temperature is from the ambient;
10. the economic optimum ΔT of a heat exchanger decreases as the average temperature of the exchange (measured for example by the LMTD), and *vice versa*;
11. avoid whenever possible the throttling of steam and gases;
12. rank the irreversibility potential of the streams: the higher the mass flow rate, the larger the possibility for losses (and for improvement!);

---

[9] We are referring specifically to codes developed along the lines of the so-called Pinch Technology [1], which constitutes by far the most successful (deterministic) tool available today to solve HEN Design problems: there are though a number of equally successful Solvers based on different approaches, which are somewhat more difficult to use and therefore not very popular among engineers. For a review, see [23].

13. use simplified exergy consumption calculations as a guide to process modifications;
14. some second Law irreversibilities can be avoided, others cannot: concentrate on those that can.

The "translation" of these rules into propositional calculus cannot be directly implemented on a one-to-one basis: therefore, we shall first try to formulate a macro-code for our Expert Assistant, and then to show how the proposed rules relate to the 13 "thermodynamic" rules of the above list. A brief explanation is added to each one of the rules presented here.

rule 2): IF (ATD > $\Delta T_{max,spec}$)
THEN (discard present configuration)

This rule avoids the waste of a thermodynamically valuable driving force caused by forcing a coupling between streams at excessively different temperature levels. $\Delta T_{max,spec}$ must be specified by the user, and is application-dependent.

rule 3): IF (TTD < $\Delta T_{min,spec}$)
THEN (discard present configuration)

This rule avoids the waste of heat exchange surface caused by forcing two streams to exchange more heat than it is technically convenient. $\Delta T_{min,spec}$ must be specified by the user, and depends on the type of heat exchanger employed.

Notice that both rule#2 and rule#3 require additional calculations, because to compute ATD or TTD for a specific coupling of streams it is necessary to solve the energy balance of the heat exchanger. Since, at least for a first run, it is possible to use average fluid properties and an average heat transfer coefficient, the computational economy resulting in the pruning of the configuration tree caused by the application of these rules is almost certainly higher than the additional computational expense, and justifies this portion of our paradigm.

rule 4): FOR (j=1, N_{cu}) **DO**:
FOR (i=1,N_h] **DO**:
IF ($T_{hot, in} > T_{pinch}$)
AND ($T_{cold, in} = T_{cu,j}$)
THEN (discard present configuration)

This rule prohibits the exchange of heat between any hot stream above the pinch and a cold utility (the *j-th* one) which is below the pinch. Therefore, application of the rule is tantamount to prohibiting heat exchange across the pinch between heat releasing process streams and cold utilities.

rule 5): FOR (k=1, N_{hu}) **DO**:

FOR (i=1,N_c] **DO**:
IF ($T_{cold, in} < T_{pinch}$)
AND ($T_{hot, in} = T_{hu,k}$)
THEN (discard present configuration)

This rule is logically symmetric to the previous one, and prohibits the exchange of heat between any cold stream below the pinch and a hot utility (the *k-th* one) which is above the pinch. Therefore, application of the rule is tantamount to prohibiting heat exchange across the pinch between heat accepting process streams and hot utilities.

rule 6): IF {$(c_p * m)_{hot} >> (c_p * m)_{cold}$}
THEN (split hot stream)

rule 7): IF {$(c_p * \Delta T * m)_{cold} >> (c_p * \Delta T * m)_{hot}$}
THEN (split cold stream)

These two rules are to be applied after the energy balance of each heat exchanger has been calculated, and avoid the coupling of two streams with very different heat capacities (a fact that is thermodynamically not correct). Notice that this is a decision taken in the realm of approximate knowledge: how much is the ">>" worth must be decided case by case by the user (we shall use the value 3 here in our example). Also, notice that the actions "split hot stream" and "split cold stream" are complex actions, that need to be undertaken by an *ad hoc* subroutine: this independent sub-program must decide how to carry out the splitting, how to re-couple the streams, and so on. Finally, notice that this is an action that goes against our goal of "minimising the number of configurations included in the search", because in reality each rule produces at least one more configuration each time it is enforced. Again, the reason for keeping the rules in our KB is heuristic in nature: HEN design experience has shown that including these rules in our procedure improves the efficiency of the HEN, by reducing overall irreversibility and producing a better (closer to "optimal") HEN configuration.

rule 8): IF (stream is hot)
AND (it has been cooled in a previous step)
AND (it is being heated up in the present step)
THEN (discard present configuration)

rule 9): IF (stream is cold)
AND (it has been heated up in a previous step)
AND (it is being cooled down in the present step)
THEN (discard present configuration)

These two rules improve the efficiency of the HEN in two ways: first, they implicitly exclude the possibility of heat transfer across the pinch; second, they avoid duplicating irreversibilities by re-heating a stream that had been cooled in a previous heat exchanger or re-cooling another that had been heated before. They have also an indirect effect of reducing the number of heat exchangers, by ensuring that a hot stream remains a heat donor and a cold stream a heat acceptor. Their enforcement though is quite resource-intensive from the computational point of view: they require memory of past actions, which in computational terms means that all previous successful nodes of the configuration tree must be kept in the working memory.

rule 10): IF (heat exchanger is of contact type)
AND {[($p_{hot} \gg p_{cold}$) **OR** ($p_{hot} \ll p_{cold}$)] **OR** ($T_{hot} \gg T_{cold}$) **OR** (chemical compositions are widely different)}
THEN (discard present configuration)

This rule avoids the mixing of streams with wide differences in pressure, temperatures or chemical composition, thus avoiding the associated irreversible losses. Notice that, as in rules 6 and 7, quantitative handling of approximate knowledge is required here.

rule 11): IF (stream is cold)
AND IF (it must be heated)
THEN (choose as heat donor the stream nearest to the prescribed exit target-temperature)

rule 12): IF (stream is hot)
AND IF (it must be cooled)
THEN (choose as heat acceptor the stream nearest to the prescribed exit target-temperature)

These two rules are symmetric, and force a "ranking" of heat donors and acceptors such that the thermal driving forces are minimised.

To check which ones among the 13 thermodynamic rules have been enforced at this level, an equivalence table can be useful:

propositional rule 2	is embedded in ($\subset$)	thermodynamic rule *a*
propositional rule 3	is embedded in ($\subset$)	thermodynamic rule *a*
propositional rules 4,5	is embedded in ($\subset$)	thermodynamic rule *c*
propositional rules 6,7	are equivalent to ($\equiv$)	thermodynamic rules *a,f*
propositional rules 8,9	contain ($\supset$)	thermodynamic rule *d*
propositional rule 10	is equivalent to ($\equiv$)	thermodynamic rule *b*
propositional rules 11,12	are equivalent to ($\equiv$)	thermodynamic rules *e,g*

From the table it is clear that rules *h,i,j,k,l* and *m* have no counterpart in our macro-code. The reason is that these rules are at a different conceptual level with respect to rules 1-12. Rules *j* and *k* belong to a meta-level, and ought to be included in a separate Inference Engine that performs the following actions:

4. ranks the streams according to their mass flow rate, and instructs the secondary Inferential Engine to consider the streams in descending rank;
5. for each successful configuration, scans the streams data base to see if some throttling would lead to an improvement over the present “optimum”, and suggests the necessary calculations, or directly performs them;

Rules *h* and *i* can, in quite an opposite fashion, be directly included in the Knowledge Base: the cost data on the Heat Exchangers could include a weighing function which relates their optimal ΔT to the LMTD. Rule *l* can be enforced procedurally in the deterministic portion of the code, the one which actually performs the calculations. Finally, rule *m* has been omitted here because it requires so much “insight” that its implementation would require yet another Inferential Engine at a meta-meta-level: even if such an ES could probably be built, we think it is more advisable to let the Expert Assistant stimulate the user’s knowledge and insight by some interactive prompting, than to embark in the difficult task of developing a “General Thermodynamic Advisor”.

The remaining functions (calculation of fluid properties, calculation of the average heat exchange coefficients, heat and mass balances, cost assessment, etc.) can be executed by any of the existing codes that perform more than satisfactorily each one in its application domain.

Our Expert Assistant (let us name it “HENEA”, short for Heat Exchanger Network Expert Assistant) is, in principle, completed: some applications to standard cases for which a solution was already known from previous independent calculations are presented in [68]. Here, let us apply it to the situation depicted in Fig. 6.22. To begin with, let us choose the numerical values of the constraints appearing in Rules 1-12; without further discussion, we shall accept the values suggested in the following table:

Constraint	**Rule where the constraint applies**	**Numerical value**
Minimum number of Heat Exchangers, N_{min}	1	5
Approach temperature Difference, ATD	2	10°K
Terminal Temperature Difference, TTD	3	50°K
Pinch Temperatures	4,5	330/340°K

Furthermore, we shall neglect the possibility of adopting contact Heat Exchangers, so that Rule 10 does not apply. Finally, since the flow heat capacities of the cold and hot streams are not too far apart from each other, stream splitting can be disregarded, and thus Rules 6 and 7 do not apply either. Notice that the Pinch Temperature must be computed in advance to enforce Rules 4 and 5: in this sense, HENEA "builds on" Pinch Technology. The procedure is straightforward: we generate a configuration tree (shown in Fig. 6.23) and apply each Rule in turn to each configuration. Since the method chosen to construct the tree is that due to Pho and Lapidus (as quoted in [23]), Rules 8 and 9 are enforced automatically. The following table shows the step-by-step results of the application of HENEA. The symbol following the numerical value of a constrained parameter indicates that the pertinent Rule is not satisfied (the value exceeds the imposed constraint).

Configuration	Rule # 1 (N_{min}=5)	Rule # 2 (ATD_{max}=10°K)	Rule # 3 (TTD_{max}=50°K)	Rule # 4 (no cold utility above pinch)	Rule # 5 (no hot utility below pinch)
A1					
A2			94.5°K		
A3					
A4			95°K		
B1			95°K		
B2			94.5°K		
B3			94.5°K		
B4	6				
B5			94.4°K		
B6					
B7			95°K		
B8			95°K		
C1			94.5°K		
C2	6		80°K		
C3	6		64.4°K		
C4					
C5	6				
C6			95°K		
D1	6		80°K		
D2	6				

The following remarks can be made:

1. the Rules are selective enough: actually, they are overselective, because no configuration passes HENEA's screening. It is clear that some of the constraints must be relaxed.
2. Rule #1 appears to be too strict: it might well be that a configuration with (N_{min}+1) Heat Exchangers is more convenient that one with N_{min}. It is perhaps advisable to reformulate this rule allowing for a derangement:

rule 1): IF ($N_{he} > \alpha N_{min}$)
THEN (discard present configuration)

3. where is a tolerance parameter of the order of 1.1-1.2. Notice that if this version of the rule had been applied, then configuration D2 would have been tagged as "optimal" by HENEA.
4. Rule #2 is always enforced, but only because the Pho and Lapidus protocol employed hee to construct the solution space allows the user to specify the ATD: so, this rule was enforced *a priori* by the designer.
5. Rules #4 and #5 are very effective in pruning the configuration tree. Recall that these rules are a direct application of one of the basic principles of Pinch Technology.
6. Rules #8 and #9 do not apply, because they are automatically enforced by the Pho-Lapidus protocol.
7. Rule #10 was intentionally excluded.
8. Rules #11 and #12 were switched off in the present application, because the configuration tree is very simple, and there are of course only 2 possible pairings for each step.
9. Finally, notice that the configuration tree has only 24 nodes instead of the 64 theoretically possible combinations indicated in Section 6.9.2: this is due to a strong pruning enacted by the Pho-Lapidus algorithm. This suggests that the performance of HENEA is sensitive to the protocol adopted to generate the feasible configurations.

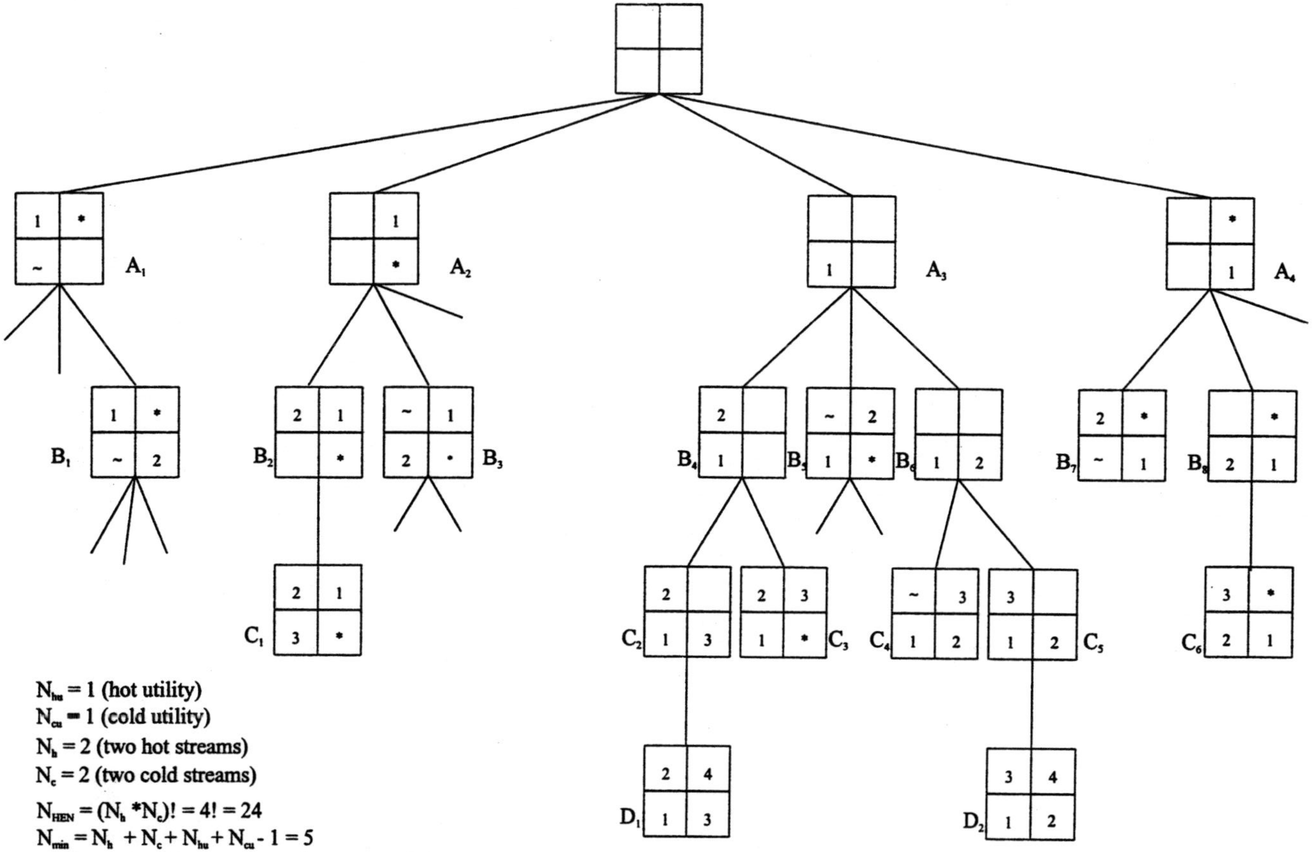

$N_{hu} = 1$ (hot utility)
$N_{cu} = 1$ (cold utility)
$N_h = 2$ (two hot streams)
$N_c = 2$ (two cold streams)

$N_{HEN} = (N_h * N_c)! = 4! = 24$
$N_{min} = N_h + N_c + N_{hu} + N_{cu} - 1 = 5$
~ = hot and cold streams coupling not possible
* = no utility necessary

Fig. 6-23 Worked-out example of the tree-generating and scanning procedure for Example 6.9

References

1. A. Bejan, G. Tsatsaronis, M.J. Moran: *Thermal Design and Optimization*, J. Wiley & Sons, 1996
2. R.A. Brooks: *Intelligence without representation*, Art.Int. 47, n.1, 1991
3. E. Charniak, D. McDemmott: *Introduction to Artificial Intelligence*, Addison-Wesley, 1985
4. N. Cross: *Engineering design Methods,* J. Wiley & Sons, New York, 1989
5. M. De Marco, M.F. Falcetta, R. Melli, B. Paoletti, E. Sciubba: *COLOMBO: an expert system for process design of thermal powerplants*, ASME-AES vol.1/10, 1993
6. D. Deutsch: *The Fabric of Reality,* Allen Lane, 1997
7. G.L. Drescher: *Made-up minds: a constructivist approach to artificial intelligence*, MIT Press, 1993
8. A. Ertas, J.C. Jones: *The Engineering Design Process*, J.Wiley & Sons, 1996
9. M.F. Falcetta, E. Sciubba: *A computational modular approach to the simulation of powerplants,* Heat Recovery Systems & CHP, vol.15, n.2, 1995
10. K.D. Forbus: *Qualitative Process Theory,* Artificial Intelligence, vol. 24, 1984
11. A.P. Fraas, M.N. Ozisik: *Heat Exchanger Design*, J. Wiley & Sons, 1965
12. C.A. Frangopoulos: *Intelligent functional approach: a method for analysis and optimal synthesis-design-operation of complex systems*, IJEEE vol. 1, n.4, 1991
13. C.A. Frangopoulos, M.R. Spakowski: *A global environomic approach for Energy Systems Analysis and Optimization,* Proc. ENSEC 93, J. Szargur ed., Cracow (Poland), 1993
14. W. Fratscher (Ed.): *Energy management for process engineer,* VEB, 1989 (in German)
15. J.A. Freeman, D.M. Skapura: *Neural Networks,* Addison-Wesley, 1992
16. R.A. Gaggioli, S. Qian, D.A. Sama: *A Common-Sense Second Law Approach for Improving Process Efficiencies*, Proc. TAIES 1989, Beijing, International Academic Publishers-Pergamon Press, 1989
17. W.M.J. Garland: *The role of Knowledge-based Systems in Heat Exchanger Selection, Design and Operation*, in "New Developments in Heat Exchangers", N.Afgan et al. (Eds.), Gordon & Breach, 1996
18. A. Gelsey: *Automated Physical Modeling*, Proc. IJCAI-89, 1989

19. J. Giarratano, G. Riley: *Expert Systems: Principles and Programming,* PWS Publ. Co., Boston, 1989
20. D. Gillies: *Artificial Intelligence and Scientific Method*, 1996
21. M. Gong, G.Wall: *On Exergetics, Economics and Optimization of Technical Processes to meet Environmental Conditions*, Proc. TAIES'97, World Publishing Corp., Beijing 1997
22. R. Gregory: *Visual Intelligence*, in "*What is Intelligence?* ", J.Khalfa ed., Cambridge Univ. Press, 1994
23. T. Gundersen, L.Näss, *The synthesis of cost optimal Heat Exchanger Networks: an industrial review of the state of the art*, Comp. Chem.Eng. v.12, n.6, 1988
24. E.C. Hohmann, F.J. Lockhart: *Optimum Heat Exchanger Network synthesis*, Proc. AIChE Meeting, Atlantic City, 1976
25. J.H. Holland: *Adaptation in natural and artificial systems*, MIT Press, 1992
26. J.J. Hopfield: *Neural networks and physical systems with emergent collective computational capabilities,* Proceedings of the National Academy of Sciences, vol. 79, 1982
27. J.P. Ignizio: *Introduction to Expert Systems*, McGraw Hill, 1991
28. D. Japikse, N.C. Baines: *Introduction to turbomachinery,* Oxford Univ. Press, 1994
29. Y. Iwasaki, H. Simon: *Causality and Model Abstraction,* Artificial Intelligence, vol. 61, n. 1, 1994
30. J.L. Kolodner: *Improving human decisions making through case-based decision aiding*, A.I.Magaz., Num. 1991
31. A.S. Kott, J.H. May, C.C. Hwang: *An autonomous Artificial Designer for thermal energy systems*, ASME Trans., Oct. 1989
32. R.A. Kowalski: *Logic for Problem Solving*, North Holland, 1979
33. B.J. Kuipers: *Qualitative reasoning: modeling and simulation with incomplete knowledge,* MIT Press, 1994
34. J.Liebowitz, D.A. De Salvo (Eds.): *Structuring Expert Systems*, Yourdon Press, 1989
35. J.W.Lloyd (Ed.): *Computational Logic*, Springer Verlag Basic Res.Series, 1990
36. M.A.Lozano,A.Valero: *Theory of the exergetic cost*, Energy, vol. 18, n.9, 1993
37. R.D. McBride, D.E. O'Leary: *The use of mathematical programming with A.I. and E.S.*, Eur.J.of Oper.Res., 70, n.1, 1993
38. J. McClelland, D. Rumelhart: *Parallel Distributed Processing,* vol.1-2, MIT Press, 1986
39. W.S. McCulloch, W. Pitts: *A logical calculus of the ideas immanent in nervous activity,* Bulletin of Mathematical Biophysics, vol. 5, 1943
40. M.L. Mavrovouniotis, G. Stephanopoulos: *Formal Order-of-Magnitude Reasoning in Process Engineering,* Computers in Chem. Eng., vol. 12, 1988
41. D.A. Mechner: *All systems GO*, The Sciences, N.Y. Acad. of Sciences, Jan./Feb. 1998

42. R.Melli,B.Paoletti,E.Sciubba: *SYSLAM: an interactive expert system approach to powerplant design and optimization*, IJEEE Jnl., 3/1992
43. M.J. Moran: *Fundamentals of exergy analysis and exergy-aided Thermal Systems Design,* Proc. NATO-ASI Neprun (Rumania), A. Bejon ed., 1998
44. M.J. Moran, H. Shapiro: *Fundamentals of Engineering Thermodynamics*, J. Wiley & Sons, 1995
45. H.G. Natke: *What is a true mathematical model? A discussion of system and model definitions*, Inverse Problems in Engineering, Vol. 1, OPA, Amsterdam 1995
46. C.V. Negoita: *Expert Systems and Fuzzy Systems*, Benjamin/Cummings Pub., 1993
47. A. Newell: *Unified theories of cognition,* Harvard Univ. Press, 1990
48. B. Paoletti, E. Sciubba: *Artificial Intelligence Applications in the Design of Thermal Systems*, in "*Developments in the Design of Thermal Systems*", R. Boehm (Ed.), Cambridge U. Press, 1997
49. P.Y. Papalambros, D.J. Wilde: *Principles of optimal design: Modeling and Computation*, Cambr. U. Press, 1988
50. K. Parsaye, M. Chignell: *Expert Systems for Experts,* John Wiley & Sons, Inc., 1988
51. J. Pearl: *Heuristics*, Addison-Wesley Pub., 1984
52. R. Penrose: *The Large, The Small, and The Human Mind,* Cambridge Univ. Press, 1997
53. W.C. Press, S.A. Teukolsky, W.T. Vetterling, B.P. Flannery: *Numerical Recipes in C*, McGraw Hill, 1985
54. R.M. Privette: *Computer-aided process design trends in the power industry,* in "*Developments in the Design of Thermal Systems*", R.Boehm (Ed.), Cambridge U. Press, 1997
55. W.V.O. Quine: *Word and Object*, MIT Press 1960
56. J. Rasmussen: *Information Processing and Human-Machine Interaction: An Approach to Cognitive Engineering,* North Holland Series in System Science and Engineering, 1986
57. G.V. Reklaitis, A. Ravindran, K.M. Ragsdell: *Engineering Optimization*, J. Wiley, 1983
58. M. Rettaroli, E. Sciubba: *MASAI: a code for the symbolic calculation of thermal plants*, ASME PD/64/3, Proc. ESDA Conf., London 1994
59. W.C. Reynolds: *Thermodynamic properties in SI,* Stanford Univ. 1979
60. E. Rich: *Artificial Intelligence*, McGraw-Hill, 1983
61. W.M. Rohsenhow, J.P. Hartnet (Eds.): *Handbook of Heat Transfer*, McGraw Hill, 1975
62. P. Roosen, B. Gross: *Optimization strategies and their application to Heat Exchanger Network Synthesis*, Proc. Eurotherm Seminar, RWTH Aachen, July 1995
63. B. Russell: *Analysis of Mind*, The Macmillan Co., 1924
64. D.A. Sama: *The Use of the Second Law of Thermodynamics in Process Design*, JERT, vol. 117, n.9, 1995

65. E. Sciubba: *Toward Automatic Process Simulators: -Part I Modular Numerical Procedures,* J. Eng. for GT and Power , vol.120, n.1, 1998
66. E. Sciubba: *Toward Automatic Process Simulators: -Part II An Expert System for Process Synthesis,* J. Eng. for GT and Power , vol.120, n.1, 1998
67. E. Sciubba: *The use of Exergy Analysis to plan for structured changes in an Industrial Society*, Intl. J. Glob. Energy Iss., 1998
68. E. Sciubba: *HENEA: an exergy-based Expert System for the synthesis and optimization of heat exchangers networks*, submitted to JERT, 1998
69. A. Scott: *The hierarchical emergence of consciousness*, Math.& Comp. in Simul., n.40, 1996
70. R. Shank, L. Birnbaum: *Augmenting Intelligence*, in "*What is Intelligence?*", J. Khalfa ed., Cambridge Univ. Press, 1994
71. C.E. Shannon, J. McCarthy, A.W. Ross: *Automata Studies,* Annals of Mathematic Studies, n.34, Princeton Univ. Press, 1956
72. W.H. Shiffmann, H.W. Geffers: *Adaptive control of dynamic systems by back-propagation networks*, Neur.Netw. 6, n.4, 1993
73. Y. Shoham, Y. Moses: *Belief as defeasible knowledge*, Proc. IJCAI-89, 1989
74. R.D. Sriram: *Intelligent Systems for Engineering*, Springer Verlag, 1997
75. G. Stephanopoulos: *Expert Systems and computing environments for Process Systems Engineering*, Comm. AIChE, n.9, 1986
76. W.F. Stoecker: *Design of Thermal Systems*, McGraw Hill, 1980
77. J. Szargut, D.R.Morris, F.R. Steward: *Exergy analysis of thermal, chemical, and metallurgical processes,* Hemisphere, 1988
78. A. Valero, L. Serra, M.A. Lozano: *Structural Theory of Thermoeconomics*, ASME-AES, vol. 2/10, 1993
79. F.J. Varela: *Toward a practice of autonomous systems*, MIT Press, 1992
80. L.E. Widman, K.A.Loparo, N.R. Nielsen: *Artificial Intelligence, Simulation and Modeling*, J.Wiley, 1989
81. L. Wittgenstein: *Philosophie*, Donzelli Pub., Roma 1996 (in Italian and German)

APPENDIX A

SHORT GLOSSARY OF ARTIFICIAL INTELLIGENCE- AND THERMAL SYSTEMS TERMS

The following list has been compiled with two goals: first, to allow Thermal Process Engineers to quickly access a clear and simple but correct source of brief explanations for the many AI terms encountered in the pages of this Book. Second, to offer the symmetric service to AI experts who would inquire about the many Process Engineering terms employed in the descriptions of the engineering problems described in Chapters 2, 4, 5 and 6. The list does not claim to be complete (some of the terms used in the Book may have escaped our attention!) nor exact (a "brief" description can not be exhaustive). But this Glossary is neither biased nor logically incorrect, that is, the definitions given here are those accepted in the current literature on AI and Thermal Engineering, and a limit has been imposed to conciseness, so that functional logical exactness would not be impaired. No sources for the information contained in this Glossary have been quoted here: it is implied that, where necessary, definitions and explanations have been extracted/adapted from the Bibliographic List (page 239). Whenever possible, a reference to the Sections of this book has been indicated.

Keyword	Section	Explanation
abstraction		Every data set can be represented by two levels of specification: a functional one, that describes what the Objects do (their function), and a practical one, that describes the actual implementation of that function (the multitude of elemental tasks that the Object performs to accomplish its function). The logical action of separating the

		functional from the practical specification level, and of taking only the former to represent the Object, is called abstraction.
aggregation level	4.1	The logical level at which components or units of a process are described. Aggregation level "0" is the one at which each single component is treated as a black-box; "1" is the level at which functional groupings are considered; ..."N" the one at which the entire process is treated as a black box.
aggregations	4.1	Super-sets or sub-sets of a system. For example, the compressor-turbine set is a sub-aggregation of a gas turbine plant. In an industrial plant, the plant is a super-aggregation of the water treatment unit, the electrical generation unit, etc.
algorithm		A finite set of clear and unambiguous elementary instructions that accomplish a particular task.
always-relevant parameters	2.4.2.2.1	In a Design Task, all those parameters that are known or thought to be of essential importance in the identification of the properties or of the operational behaviour of the system to be designed.
analogy	4.1; 5.1; 5.2	A particular type of logical link between two Objects whose attributes may not be the same nor in the same "order", but are mappable onto each other at some meta-level: this means that there exists *one* set of rules capable of representing the relationships among corresponding relevant attributes of both objects.
AND-tree	3.1.16; 3.9.2	A decision tree that possesses only AND nodes
AND/OR tree	3.1.16; 3.9.2; 3.9.3	A decision tree consisting of at least one AND/OR node
approximate reasoning	3.8	Logical deduction based on approximate knowledge, i.e., on a knowledge base expressed by non-exact facts. *Approximate* KB ought not to be confused with *fuzzy* KB: a characteristic of approximate statements is that they can always somehow (albeit at times in an awkward format) be expanded in

		such a way that they give origin to "exact" expressions.
Artificial Intelligence	3.1.1	The portion of Computer Science that investigates symbolic and non-algorithmic reasoning, and attempts to represent knowledge in such a form that it can be employed to generate instances of machine inference.
attribute		A property of an Object. Said to be *characteristic* if it is the attribute on which the belonging of the Object to a Class is estabished.
automatic operation of a plant	6.3	The implementation of a series of monitoring- and control devices connected to a central unit that co-ordinates the flux of information from- and to the processes and manages them according to either a predetermined plan ("non-intelligent", or rigid, control) or to a general schedule which attempts to meet some general goals ("intelligent" or flexible control).
backward chaining, BC	3.1.12	A form of reasoning that attempts to validate (i.e., confirm) a fact (called "goal") by scanning the Knowledge Base to determine whether the rules therein contained univocally generate the logical possibility of the goal. In practice, this means that if a fact **q** is asserted by the KB, every rule that contains **q** as the "THEN"-predicate is assumed to be true, and therefore **p**, the predicate of its "IF", is also true.
backtracking	3.11	When a scanning procedure reaches a "dead branch" (a node N_{ij} which does not possess suitable successors) in the tree it is exploring, it must resume the search from some other node N_{hk}: if $h=i-1$, $k=j$, i.e., N_{hk} is N_{ij}'s predecessor, the backtracking is called *chronological*; otherwise, it is *procedure-driven* (for instance, breadth search resumes from $N_{i,j-1}$, depth search from $N_{i-1,j}$, etc.). Backtracking can also be directed by a logical-connection list, and in this case it is called *dependency-directed*.
belief	3.8	A particular form of approximate

		knowledge: a fact is asserted not with absolute or however quantifiable certainty, but with a vague and non-quantifiable proposition: "I believe that *f* is true". Belief can be quantified by assigning a computable degree of unlikeliness to any counterfact that would disprove the believed fact.
blackboard system	3.1.15	A particular form of hierarchically organised ES, consisting of a certain number of local ES ("sources"), each handling a portion of the KB and communicating their conclusions to a "Blackboard" which represents the "latest state of affairs" of the global inferential activity. A dedicated IE controls the flow of information and resolves contradictions among different sources.
Boiler		A device used to thermally evaporate a flow of water, transforming it into steam, and then to superheat it, by means of heat transfer from a flame obtained by burning a proper fuel. In modern industrial boilers, the boiling water at high pressure (30-220 bars) flows through "firewalls" of tubes, welded together to increase the heat exchange surface. The superheated steam exits the Boiler at 550-570° C.
certainty		An event is said to be *certain* when the probability of its negation is unconditionally equal to zero. Engineering data are, for their vast majority, not absolutely certain: this can be accounted for by attaching to the corresponding propositions a certainty factor $0 \leq f_c \leq 1$, and applying the rules of approximate reasoning.
class		A collection of Objects, based on some common attribute of its components, or on some peculiar relationship among them. Notice that there may be more characteristic attributes and multiple characteristic relationships.
clause		The predicate of a logical connective (IF and THEN) in a rule
Combustion Chamber	2.4.1	A device in which air is mixed with a fuel and the mixture is then properly ignited. The

		resulting oxidation reactions of the fuel molecules (the net sum of which constitutes the Higher Heating Value of that fuel) provide the energy necessary to raise the temperature of the air/combustion products mixture to very high temperatures (1700-2500 °C, depending on fuel and fuel/air ratio)
Compressor	2.4.1; 6.1	A device used to pressurise a gas flow.
compressor stall	6.1	An undesirable operational situation in which fluid dynamic instabilities on the Compressor blades lead to pressure fluctuations on the outflow and possibly to flow-induced vibrations on the structure. If not corrected, stall leads to *surge*, a more severe pathological compressor behaviour.
Condenser		A device used to condense a flow of steam, subtracting its latent heat of condensation by means of a flow of water or air. The pressure and the temperature in the Condenser are related by the equilibrium relations, because the condensed water is at the saturation pressure corresponding to its temperature.
connectivity graph	2.4.1.5	A graph consisting of points (nodes) each representing an Object (or event), connected by lines which represent the interconnections between pairs of nodes. These interactions may be logical or physical, and may possess an implicit "direction" (i.e., be equivalent to an arrow) or be direction indifferent. Sometimes, each connection bears a "cost label", which identifies the price (monetary, energetic, or functional) which one incurs in if that route is taken.
constraint	3.10	Any restriction posed by technological, environmental, legal or practical requirements on the configuration of a system. Constraints may limit the values of some process parameter, or exclude some system configuration.
criteria for success	2.1.1.2	A set of attributes of the Object to be designed that can be assigned quantitative values so that the probability of a particular design to be successful can be assessed.

data		A collection of facts, relationships and propositions about a portion of the Universe.
data type		A collection of Objects and of operations to be performed on these Objects.
Deaerator		A Mixing Heat Exchanger in which “hot” steam is insufflated into “cold” water to bring the liquid to its boiling point, so that all uncondensable gases are released. These gases are then captured by proper “gas traps”, and the deaerated water is pumped to the Boiler.
decision tree	3.1.16	A graphical representation of the set of possible choices for a given situation. The “tree” can be thought of as a single trunk giving origin to several branches, each branch having twigs, each twig bearing twiglets... until the leaves (or fruits) are reached.
declarative knowledge	3.1.2	Knowledge of facts and relationships expressed in propositional form: “Fact **A** is an Object **B** having the relationship **p** with the Object **C**”. Also called “*knowing what*”.
deduction	3.1.11	A logic procedure that marches forward from the causes to their effects.
deep knowledge		Substantial knowledge of a problem, which includes a complete understanding of the premises, a “feel” for the outcome and a vast experience of application cases, that implies a qualitative and quantitative comprehension of the logical chain of reasoning and of the underlying physical phenomena. An individual possessing deep knowledge about a field is called an *expert* of that field.
degree of membership	3.5	The degree to which a value belongs to a fuzzy set. For example, a level measurement may be 40% positive large and 20% positive medium.
deterministic programming	3.2	Numerical methods stemming from the quantitative modellization of a portion of the Universe and the discretization of the model equations. They usually assume that at least one solution exists for the problem under consideration, and generate solutions that are

		strongly model-dependent.
direct design problem	2.3	A problem which requires the Engineer to find the "best" solution to attain a specified design objective for an assigned process configuration.
Domain Expert	3.1.18	An expert in a specific field related to the domain of an ES, whose knowledge must be imparted to the ES.
embedding	5.1	"Containing", in logical sense. A procedure **p** is said to embed another procedure **q** if every time **p** is performed, **q** is too, but not vice-versa. Notice that **p** is at a meta-level with respect to **q**.
emission monitoring	6.2	A series of procedures by which the effluents from a process are thermally and chemically analysed continuously or at prescribed intervals in time, so that the relevant values derived by these measurements can be compared with the values specified by applicable regulations.
encapsulation		The concealing of the details of the implementation of an Object. For example, a black-box approach encapsulates the actual internal structure of the component it represents, and projects to the outside user only the function performed by that component.
Exergoeconomics	2.6.2	Commonly defined as the branch of Thermoeconomics that uses (monetary) costing functions calculated using exergetic criteria. Here proposed with a different meaning: the branch of Economics that uses exergy-, and not monetary, costs in its accounting procedures.
facts	3.1.7	Expressions that describe particular situations. The expressions can be logical (propositions), symbolic ("**p=q**"), numerical or probabilistic.
Feasibility Study	2.1.1.6	One of the sub-task of any design activity. It consists in creating a preliminary concept configuration for the process, approximately sizing the main equipment, and executing a preliminary technological- and cost analysis.
feedback		In the context of this book, this word has two

		meanings: 1) a control system is said to possess a *feedback mechanism* when its output (i.e., the controlling action) depends on the value assumed by the controlled quantity downstream of the control; 2) a process is said to have *feedback fluxes* if - seen from the point of view of its main product- the process diagram is not linear, because certain components receive inflows from downstream (the P&I displays some "feedback loops"). See also "logical loop".
Feedwater Pump	6.4	A device used to pressurise the water coming from the Condenser so that it attains the same pressure required by the Boiler, plus friction losses. Usually a multi-stage centrifugal pump. In large Steam Plants, the Feedwater Pump is located immediately downstream (and underneath) of the Deaerator, and upstream of the high-pressure water Preheaters.
forward chaining, FC	3.1.13	A form of reasoning that attempts to deduce a hitherto "unknown" fact (called "goal") by scanning the Knowledge Base to determine whether the data therein contained may generate the logical possibility of the goal. In practice, if a fact **p** asserted by the KB is the "IF"-predicate of a rule, then we can infer that **q**, the "THEN"-predicate of that rule, is also true.
frame		A collection of Semantic Network nodes and links. A frame is an Object denoted by its name, endowed with some slots, each slot having more facets: slots and facets may store values, attributes or relations.
fuzzy knowledge		A synonym for vague knowledge. Vagueness may be introduced into a problem by incomplete or incorrect data, incorrectly expressed relationships, or by the use of an inappropriate model.
fuzzy search	3.9.4	A tree-scanning procedure in which the searching criterion is fuzzified
fuzzification	3.5	The process of decomposing a correlated set of knowledge chunks (usually available in an uncertain or vague fashion) into one or more

		qualitative groups called fuzzy sets.
fuzzy set	3.5	A set admitting a non-binary degree of membership. In "crisp" set theory, an Object either is or it is not a member of a certain set. In Fuzzy Set Theory, a degree of membership $0< d_m <1$ can be attached to each Object and express the likelihood of its belonging to a given set: d_m is clearly a particular kind of relationship between a set and its Objects. "Normal" language expressions ("very likely", "highly improbable", "almost certain", "more likely than") can be easily quantified by a fuzzy approach: therefore, FST is well suited to handle vague knowledge.
game analogy	5.6	A logical similarity is easily established between ES's and games. For Design-oriented ES, the mapping produced by this analogy is based on the following projections: Game Hardware & Rules → Problem Data; Game Strategy → Design Rules; Game Victory → Design Objective. Consequently, many of the existing game-related tree-searching techniques can be converted into Design Solution Strategies.
Gas Turbine	2.4.1	A device through which a hot pressurised gas is expanded to yield shaft power.
Heat Exchanger	6.5	A device used to transfer a certain amount of heat from a "hot" fluid to a "cold" one. This exchange can take place indirectly, through a metallic wall (shell-and-tube HE, cross-flow HE, compact HE, radiators, etc.); or by mixing (deaerators, attemperators). In the first case, the HE is said to be a contact HE, in the second case, a mixing HE.
Heat Exchanger Network	6.9	A series of HE connected in several "layers" so that they form a matrix of physical devices through which several fluids exchange heat. A HEN cannot be always so designed that its net energy balance (algebraic sum of heat given or taken by each single flux) closes to zero: therefore, a hot and a cold utility (external heat sources)

		are usually inserted to compensate.
heuristics		A set of rules, often approximate and vague, dictated by intuition, experience and judgment. Can be also seen as a consistent but incomplete set of propositions about a chunk of specific knowledge for which an exact description is either unavailable or impossible. Proper application of heuristics, especially in the early stages of a search process, drastically limits the solution space.
hierarchical refinement	3.1.14	Top-down procedure by which a Design Plan is implemented in descending levels of abstraction: first the general Plan Layout, then a detailed Preliminary Plan, then a general Activity Schedule, then a detailed Activity Schedule, etc.. HR is though a more general concept, that has useful applications in several fields, for example in Process Synthesis.
Hydraulic Turbine		A device in which a liquid (usually, water) looses a portion of its total head and confers the equivalent energy to a rotor which then drives a mechanical shaft.
ill-posed		A not well-posed problem
ill-structured	2.1	A not well-structured problem
induction	3.1.10	A logic procedure that marches backwards from the effects to their causes
inference		The derivation of conclusions from premises. Logically expressed in propositional calculus by "IF [(**a** AND **b**) OR **c**] THEN **d**" rules.
Inference Engine, IE	2.4.2.1; 3.1.5	A consistent and ordered set of rules, embodied in a shell-like program that establishes, controls and instantiates the problem-solving strategy of an ES
Input Plant Matrix, IPM	2.4.1.5	A process-dependent structured data-table that contains in a predetermined sequence all the relevant information necessary to launch a Process Simulation software.
instance of a class		A specific Object of the class, identified by a quantitative or qualitative value assigned to one of its characteristic attributes.
Interconnection Matrix, IM	2.4.1.5	A process-dependent structured data-table containing information about the connectivity of a Process. IM is in fact the

		matricial representation of the connectivity graph of the Process.
inverse design problem	2.3	A problem which requires the Engineer to design a device or a plant given the design objective and an objective function that must be extremized in the course of the solution.
knowledge	3.1.2	A qualitative and/or quantitative description of a specific portion of the Universe, in the form of a collection of data, i.e., of certain sets of facts, relationships, rules, forecasts, estimates and numerical expressions.
knowledge acquisition	4.2	The collection and ordering of knowledge. Requires some knowledge at a higher level than the problem being considered (meta-knowledge).
Knowledge Base, KB	3.1.4	All the knowledge collected by the knowledge acquisition activity.
knowledge decomposition	4.2	The subdivision of the KB into elemental information bits, i.e., into the largest possible set of the smallest possible data. KD forms the conceptual basis of Semantic Networks.
Knowledge Engineer	3.1.17	A person dedicated to the eliciting, gathering and collecting of knowledge (typically, from Domain Experts), and to its organisation into a KB
knowledge representation	4.1	Organisation of the "raw" KB (i.e., "as collected" knowledge) into a coherent, comprehensive and workable KB.
language	5.4	A logical system whose semantics are based on some facts representing our knowledge of a portion of the Universe. In other words, the symbols of a language always mean something "real" (an Object of the real world, an abstract Object of an imaginary world, a concept accepted by its own use, etc.).
logical loop		A chain of formal propositions $p \rightarrow q \rightarrow r \rightarrow \ldots \rightarrow z$ whose last term z is the premise for the first one, r. To avoid being a tautology, a logical loop must be implemented on *instances* of propositions, i.e., on facts: in this case, it means that there is a feedback mechanism, driven by the output of the last fact (component, flux) z, which influences

		the input of the first fact (component, flux) *r*.
logical system	3.3	A system consisting of symbols ("signs", "words") and of a complete and coherent set of rules ("grammar", "syntax", "semantic") that describe the correct use of the symbols.
memory	3.7	The ability of an ES to "remember" facts not originally contained in its KB, but constructed or deduced or induced in the course of the inference process.
macro-structuring	5.1	The process of collecting under a single ES solution procedures for different problems. Usually, the macro-structure consists of a macro-IE that drives several specific IE's, one for each problem (as in Blackboard Systems, for example).
membership function	3.5	The curve attached to a fuzzy set which maps an input (or output) value onto a corresponding value for the degree of membership.
meta-structuring	5.1	The process of devising a single higher-level procedure capable of solving distinct, but logically similar problems. The meta-procedure is likely to be simpler than each of its instances.
modular simulator		A Process Simulator that is not dependent on the structure of the process, but can be applied to whatever structure one can build using the (modular) Components and Fluxes contained in its Library.
neural network	3.6	A set of nodes (neurones) arranged in layers, each neurone of each layer connected to all of the neurones of the previous ("upper") and following ("lower") layer, but not to the neurones of its own layer. Neurones communicate messages to each other, and the firing of a message depends on the state of the neurone at the particular time of firing. If the lowest level experiences some environmental change (it is given some input), the information propagates until it reaches the uppermost level: the modification of the states of the neurones of this uppermost layer, taken in its globality,

		represents the answer of the NN to the initial stimulus.
objective function		The formalisation (not necessarily in numerical form) of the criteria for optimality of a solution
Objects	3.1.8	A computational structure that contains data, data structures and related procedures. This means that an Object can operate on itself, because it contains both the data and the operational information on how to process them.
optimisation	2.3, 2.5, 2.6	The mathematical procedure that leads to the quantitative minimisation or maximisation of a certain expression of the state parameters of the process under consideration, called the procedure's *objective function*; the set containing the values of the state parameters that extremise the objective functions are called the optimal solution set. In qualitative terms, the construction of the "best possible" solution to a given problem.
OR-tree	3.1.16; 3.9.3	A decision tree that possesses only OR nodes
Output Plant Matrix	2.4.1.5	A process-dependent structured data-table that contains in a predetermined sequence all the relevant information necessary to completely express the results of a Process Simulation software.
paradigm		An underlying principle of organization that constitutes the conceptual basis for a programming language or a solution procedure.
possibly-relevant parameters	2.4.2.2.2	State parameters of a process that are not essential to its description, but may become essential as a consequence of external circumstances (imposition of constraints, activation of a correlation with other, relevant, parameters, etc.)
predicate calculus	3.3	A pseudo-language consisting of logical predicates (facts describing Objects and their relationships) joined by a small and fixed set of connectives (AND, OR, IF..THEN, EQUAL TO, LARGER THAN, FOR ALL...THEN) that establish mutual correlations between predicates. The only

		form discussed in this book is the first order predicate calculus, in which Objects may assume quantitative values but predicates cannot.
primitive problem	2.2.1	The form in which an Engineering problem is first formulated: it is usually ill-structured, i.e., vague, incomplete, with a non-precisely formulated objective function and with unpredictable knowledge about its solution.
probability of success	2.1.1.3	Qualitative or quantitative likeliness that a product is produced within schedule and budget, meets the objectives and corresponds to the criteria for success pre-set for it.
procedural knowledge	3.1.2	Knowledge of the inference procedure to apply to a certain data set to reach certain conclusions. Also called "*knowing how*".
procedure		A set of sub-tasks with a pre-determined interconnection, whose correct execution leads to the fulfillment of a task or project.
process		A series of transformations that modify the physical state of some specified amount of matter.
process simulator	2.4.1	A computational software that deterministically performs all the calculations necessary to compute the mass- and energy balances of a process.
production rule		Synonim for rule.
propositional calculus	3.3	A pseudo-language consisting of logical propositions (assertions about Objects and their relationships) joined by the four fundamental connectives (AND, OR, IF..THEN, EQUAL TO) that establish mutual correlations between these propositions. Propositional calculus is not quantifiable: it allows though to draw logically necessary conclusions from a set of premises.
property	4.1	Attribute of an Object.
prototype		A working instance of a computational code, usually of reduced complexity than the finished product but having all of its essential features (specifically, the IE). Generally, the prototype version has the same problem-solving depth of the final

		version, but much less breadth, i.e., it can handle a much smaller solution space.
Pump		A device in which a liquid (usually, water) increases its total pressure by interacting with a rotor driven by a mechanical shaft.
qualitative reasoning	5.2	A form of reasoning used in process modelling. It is based on non-quantitative concepts, like: analogy, similarity, asymptotical behaviour, interactions, order-of-magnitude values, relationships, structure, functionality.
recursive algorithm		An algorithm A that calls upon itself during its own execution. Recursion can be *direct*, if the link is of the type "call (A)"; or *indirect*, if the link is of the type "call (B)" and B contains a "call (A)" statement.
relational programming	3.2	Non-numerical methods resulting from the qualitative modellization of a portion of the Universe and the conceptual representation of the context in which the problem arises. They do not assume *a priori* that a solution exists, and generate models for the solutions rather than the solution themselves.
relationships	4.1	Connections between Objects.
rules	3.1.6	Logical construct of the form "IF **p** THEN **q**", where both **p** and **q** are propositions or instances of propositions.
search methods	3.9	Numerical or qualitative techniques that scan a certain number of alternatives to find the more desirable one. The search can be performed on layered structures (decision trees), in which case the search criterion may be local ("Among the successors of a node, proceed to the one with the most convenient value of a certain state parameter") or global ("Choose the path composed by the nodes and branches such that the overall value of a certain function of both a state and a path parameter is the most convenient"). Global searches are obviously more difficult, and special methods have been devised to limit the number of nodes explored by the scanning procedure.
Selective	2.4.1.7	A numerical solution technique for the

Segregated Solver, SSS		system of equations representing a certain model of a process. The SSS scans the matrix of coefficients of the system to be solved, and selects (solves) only those equations that present one unknown. The calculated values are then used in a subsequent scanning, so that other equations now can be solved, and the process is iteratively repeated until either a solution is reached or the "Selection" fails, because there are equations with more than one unknown. In this last case, the problem is ill-posed (specifically, underspecified), and additional data must be supplied.
semantic network	3.4	A set of nodes and of their connections: each node represents an Object, and each connection expresses a formal relationship between Objects, chosen in a list of "allowable" relationships.
shells	5.5	A domain independent framework of rules connected to form a very general Inference Engine with specific problem solving capabilities. A shell is used by implementing its "reasoning" on a set of specific (i.e., domain- and problem-oriented) knowledge added modularly to the KB of the shell.
Steam Turbine	2.4.1	A device through which a hot pressurised vapour (usually, water steam) is expanded to yield shaft power.
symbolic reasoning		The application of manipulation rules (*syntax*) to a set of pre-defined symbols. SR follows the rules of predicate calculus.
taxonomy		A systematic set of rules used for classification purposes. Given a data-base composed of various chunks of knowledge and raw data, knowing its taxonomy means, for example, being able to partition the data-base into distinct but correlated subsets in each of which a different IE may be applied.
Thermal System		A collection of orderly interconnected devices/components whose goal is that of operating on an input supplied under the form of thermal energy.
Thermoeconomics	2.6.1	Second-Law based cost optimisation

		technique. The costs (in monetary units) are computed with the aid of entropic- and exergetic considerations.
time-marching	2.4.1.8	A numerical integration technique, based on the discretization of the time axis. Given the value of a certain function f at $t=t_o$, its value at $t= t_o + \Delta t$ is given by: $f(t_o + \Delta t) = f(t) + M \langle t_o, f(t_o), \Delta t \rangle$, where M is the so-called time-marching operator.
transfer function	2.4.1.1	A logical, symbolic or mathematical operator expressing the functional relationship between the inputs **I** and the outputs **O** of a process or component: $\mathbf{O} = \Pi(\mathbf{I},\mathbf{O}) \langle \mathbf{I} \rangle$.
weighted search	3.9.2	A search process in which each branch of the decision tree carries a "weight" or "penalty" function: the objective of the search is that of finding the path(s) for which the sum of the weights on the n branches constituting the path(s) is minimal.
well-posed		In a strict mathematical sense, a problem is said to be well-posed if it admits of one and unique solution, and this solution depends continuously on the relevant data of the problem (boundary and initial conditions, values of the coefficients, etc.).
well-structured	2.1	A problem that can be described in terms of numerical variables, possesses a univocally and well-defined objective function, and admits of a algorithmic routines for its solution.

APPENDIX B

BRIEF LIST OF COMMERCIALLY AVAILABLE AI-SHELLS

Note: the following list is included here solely as a help to the interested reader, and is intended to be a first-level help in the selection process. No commercial intent guided its compilation, the data being taken from open sources (namely, INTERNET!). Also, no guarantee is given that the shells actually perform as represented in the documentation issued by the Owner. If there are omissions or errors, the Authors will be grateful to those who will detect and correct them.

1. ACQUIRE

Characteristics: Full development environment for building and maintaining KB applications. Utilities for knowledge acquisition, rules representation, uncertainty handling, qualitative description.
Platforms: unknown
Commercial contact: Acquire Intelligence, Inc., Suite 205
McKenzie Ave.,
Victoria, Canada V8P 2L5
Ph: (604)479-8696 Fax: (604)479-0764
e-mail: sales@aiinc.bc.ca ; URL: http://vvv.com/ai/

2. ACTIVATION FRAMEWORK

Characteristics: Tool for real-time data interpretation applications
Platforms: DOS, Windows, UNIX
Commercial contact: The Real-Time Intelligent Systems Corporation
Worthen, Chelmford, MA 01824

Ph: (508)250-4633 Fax: (508)256-8132
e-mail: rtis@world.std.com

3. Aion Development System (ADS)

Characteristics: Object-Oriented knowledge representation, forward- and backward chaining engine
Platforms: DOS, Windows, UNIX, OS/2, VMS
Commercial contact: Aion Corporation
University Ave.
Palo Alto, CA 94301
Ph: (415)328-9595

4. Angos Knowledge Seeker

Characteristics: Data mining tool
Platforms: DOS, Window
Commercial contact: Angos Software International Ltd.
King Street W, Suite 201
Toronto MSU 1J5 Canada
Ph: (416)593-1122 Fax: (416)593-5077

5. Art*Enterprise

Characteristics: Object-Oriented tool, forward chaining engine (backward chaining can be implemented). Data mining tool
Platforms: Windows, UNIX, OS/2
Commercial contact: Brightware Inc.
Rowland Way, suite 310
Novato, CA 94945
Ph: (415)899-9070 Fax: (415)899-9080
e-mail: info@brightware.com ;
URL: http://www.brightware.com

6. Arity

Characteristics: Rule-based and frame-based knowledge representation
Platforms: DOS, OS/2
Commercial contact: Arity Corporation
Damonmill Sqaure
Concord, MA 01742
Ph: (508)371-1243 Fax: (508)371-1487
e-mail: 73677.2614@compuserve.com

7. Clips

Characteristics: C Language Integrated Production System. Developed by NASA Johnson Space Center, is an Object-Oriented production system language implemented in C which utilizes the RETE algorithm for pattern matching
Platforms: DOS, Windows, Unix
Commercial contact: GHG
1100 Hercules, Suite 290
Houston, TX 77058
Ph: (231)488-8806 Fax: (231)488-1838
e-mail: info@ghg.net; URL: http://www.ghg.net

8. Comdale

Characteristics: Object Oriented real-time expert system for industrial process monitoring
Platforms: DOS
Commercial contact: Comdale Technologies Inc.,
The Comdale Building
Evans Ave., Suite 600
Toronto, Ontario, Canada M9C 1A3
Ph: (416)620-1234 Fax: (416)620-4526
e-mail: info@comdale.com;
URL: http://www.comdale.com

9. Easy Reasoner

Characteristics: Adaptive associative reasoning, retrieves cases for similarity from databases. Multiple decision.
Platforms: Windows, UNIX
Commercial contact: The Haley Enterprise Inc.
Orchard Street
Sewickley, PA 15143
Ph: (412)741-6420 Fax: (412)741-6457
e-mail: info@haley.com ;
URL: http://www.haley.com

10. Eclipse

Characteristics: Derived from ART and compatible with CLIPS. Data driven forward and backward chaining. Database integrate.
Platforms: DOS, Windows, UNIX
Commercial contact: The Haley Enterprise Inc.
Orchard Street
Sewickley, PA 15143
Ph: (412)741-6420 Fax: (412)741-6457
e-mail: info@haley.com ;
URL: http://www.haley.com

11. Emerald Empower Procedural Advisor

Characteristics: Tool to develop trouble-shooting expert systems.
Platforms: MacIntosh, Windows
Commercial contact: Emerald Intelligence, Inc.
3850E Research Park Dr.
Ann Arbor, MI 48108
Ph: (313)663-8757 Fax: (313)663-8757

12. Exsys Professional

Characteristics: Backward and forward chaining, fuzzy logic, neural networks. Interface to Databases.
Platforms: DOS, Windows, MacIntosh, UNIX
Commercial contact: Exsys Inc.
Lousiana Boulevard NE, Suite312

Albuquerque, NM 87110
Ph: (505)256-8356 Fax: (505)256-8359
e-mail: info@exsysinfo.com ;
URL: http://www.exsysinfo.com

13. Flex

Characteristics: Forward and backward chaining; frames, rules and questions are described in an English-like language. Interface to several PROLOG languages including Quintus and BIM
Platforms: Windows, UNIX
Commercial contact: Logic Programming Associates Ltd.
Studio 4 RVPB
Trinity Rd.,
London, SW18 35X
Ph: 181-871-2016 Fax: 181-874-0449
e-mail: lpa@cix.compulink.co.uk ;
URL: http://www.lpa.co.uk

14. Gensym - G2

Characteristics: Tool for dynamic real-time applications. Concurrent execution of rules and procedures. Interface to relational databases
Platforms: Windows, UNIX
Commercial contact: Gensym Corp.
Cambridge Park Dr
Cambridge, MA 02140
Ph: (617)547-2500 Fax: (617)547-1962
e-mail: info@gensym.com ;
URL: http://www.gensym.com

15. GBB - Generic Blackboard Framework

Characteristics: Blackboard-based
Platforms: DOS, Windows, Macintosh, UNIX
Commercial contact: Blackboard Technology Group, Inc.
Main Street
Amherst, MA 01002
Ph: (413)256-899 Fax: (413)256-3179
e-mail: gbb-user-request@bn.csumass.edu ;
URL: http://www.bbtech.com

16. GURU

Characteristics: Forward and backward chaining, fuzzy logic. Interface to relational databases
Platforms: DOS, Windows
Commercial contact: Micro Data Base Systems, Inc.
Cumberland Avenue,
Box 2438 West Lafayette, IN 47906-2438
Ph: (317)463-7200 Fax: (317)463-1234
e-mail: info@mdbs.com

17. HUGIN

Characteristics: Tool for expert systems development with uncertainty handling
Platforms: Windows, UNIX
Commercial contact: Hugin Expert A/S
Niels Jernes Vej 10,
DK-9220 Aalborg
Ph: +45 9815-6644 Fax: +45 9815 8550
e-mail: info@hugin.dk ; URL: http;//hugin.dk

18. ILOG

Characteristics: Object-Oriented forward chaining tool. Non-monotonic reasoning.
Platforms: DOS, Windows, OS/2, UNIX
Commercial contact: ILOG Inc.
Landing Drive
Mountain View, CA 94043
Ph: (415)390-9000 Fax: (415)390-0946
e-mail: info@ilog.com ; URL: http://www.ilog.fr

19. ILOG RULES

Characteristics: Forward chaining engine
Platforms: DOS, Windows, OS/2, UNIX
Commercial contact: ILOG SA, 9
Rue de Verdun, BP 85
Gentilly Cedex, France

Ph: +33 (1) 49-08-35-00 Fax: +33 (1) 49-08-35-10
e-mail: info@ilog.fr ; URL: http://www.ilog.fr

20. KEE

Characteristics: ATM, rule-based reasoning, Object-Oriented
Platforms: DOS, Windows, INIX, Lisp Machine
Commercial contact: IntelliCorp, Inc.
El Camino Real West, Suite 101
Mountain View, CA 94040-2216
Ph: (415)965-5700/5500 Fax: (415)965-5647

21. KNOWLEDGE WORKS

Characteristics: Object-Oriented forward chaining, user-defined conflict resolution, database interface
Platforms: UNIX
Commercial contact: Harlequin Ltd.
Barrington Hall
Barrington, Cambridge, CB2 5RG
Ph: +44 1223-873-800 Fax: +44 1223-872-519
e-mail: knowledgeworks-request@harlequin.co.uk

22. M.4

Characteriistics: Backward and forward chaining, pattern matching, uncertainty handling.
Platforms: DOS, Windows, Macintosh, UNIX
Commercial contact: Teknowledge Co.
Embarcadero Road
Palo Alto, CA 94303
Ph:(415)424-0500 Fax: (415)493-2645
URL: http://www.teknowledge.com/M4

23. NEURON DATA

Characteriistics: Rule/knowledge based expert system development in distributed, Object-Oriented environment. Applications are rule-driven when they contain a large number of rules and procedures. Forward and backward chaining

Platforms: Windows, UNIX

Commercial contact: Neuron Data
1310 Villa Street
Mountain View, CA 94041
Ph:(650)528-3450 Fax: (650)943-2752
e-mail: info@neurondata.com;
URL: http://www.neurondata.com

24. RAL

Characteristics: High performance tool; full integration with rules and Objects written in C. Also available real-time features.

Platforms: DOS, Windows, OS/2, UNIX

Commercial contact: Production Systems Technologies, Inc.
Baum Blvd.,
Pittsburgh, PA 15213
Ph: (412)683-4000 Fax: (412)683-6347

25. RETE++

Characteristics: Forward and backward chaining; uses C++ Object-Oriented environment to create rule-based syntax and classes and data instances. Also available with add-on modules for case-based reasoning and integration with databases.

Platforms: Windows, UNIX

Commercial contact: The Haley Enterprise, Inc.
Orchard St.
Sewickley, PA 15143
Ph: (412)741-6420 Fax: (412)741-6457
e-mail: info@haley.com ;
URL: http://www.haley.com

26. VBXPERT

Characteristics: VBX extends rule-based knowledge and expert system capabilty to Visual Basic Applications by encapsulating the Rete algorithm

Platforms: Windows

Commercial contact: The Haley Enterprise, Inc.
Orchard St.

Sewickley, PA 15143
Ph: (412)741-6420 Fax: (412)741-6457
e-mail: info@haley.com ; URL: http://www.haley.com

APPENDIX C

LIST OF SYMBOLS AND ACRONYMS

a) Symbols related to Thermal Systems Engineering

• C	Component
• c_p	Specific Heat at constant pressure
• CO	Carbon Monoxide
• D_{le}	Inlet external diameter
• D_{li}	Inlet internal diameter
• D_2	Exit diameter
• $DeSO_x$	Desulphurisation Unit
• FBC	Fluidised Bed Combustor
• h	enthalpy
• H	hydraulic head
• HE	Heat Exchanger
• HEN	Heat Exchanger Network
• HHV	Higher Heating Value
• I	Input
• k	isoentropic exponent
• °K	Degrees Kelvin
• kg	kilogram (mass)
• kJ	kilo-Joule
• kWh	kilo-Watthour
• LHV	Lower Heating Value
• LMTD	Logarithmetric Mean Temperature Difference
• m	mass flow rate
• MPa	Mega Pascal
• MW	Megawatt
• NPSH	Net Positive Suction Head (pump cavitation parameter)
• O	Output
• p	pressure
• P_{el}	Electrical Power
• P_{motor}	Mechanical Power
• P_{pn}	Pumping Power
• P_{shaft}	Mechanical Power (at a shaft)

- P_{therm} Thermal Power
- Q Heat Flux
- rpm revolutions per minute
- S Sulphur
- SO_2 Sulphur Dioxide
- SSS Selective Segregated Solver
- t time
- T temperature
- TTD Terminal Temperature Difference
- U Rotational velocity
- V Absolute velocity
- w work
- x quality to a two-phase mixture

b) Symbols related to Artificial Intelligence

- ADS Alternate Direction Search
- AI Artificial Intelligence
- ANN Artificial Neural Network
- ATD Approach Temperature Difference
- BC Backward Chaining
- ES Expert System
- FC Forward Chaining
- FST Fuzzy Set Theory
- IE Inference Engine
- KB Knowledge Base
- KBS Knowledge-Based Systems

c) Other Acronyms

- IM Interconnection Matrix
- IPM Input Plant Matrix
- LHS Left-hand side
- OPM Output Plant Matrix
- Q&A Quality Assurance
- R&D Research and Development
- RHS Right-hand side

d) Greek Symbols

- α fuel-to-air ratio
- θ under-relaxation parameter
- η efficiency
- κ isoentropic factor
- Π Input-Output transfer matrix

- φ flow coefficient
- Δp pressure differential
- δt time interval
- ΔT temperature difference
- ψ head coefficient

INDEX

T

V

W